Functional Analysis
and
Linear Control Theory

This is Volume 156 in
MATHEMATICS IN SCIENCE AND ENGINEERING
A Series of Monographs and Textbooks
Edited by RICHARD BELLMAN, *University of Southern California*

The complete listing of books in this series is available from the Publisher
upon request.

Functional Analysis
and
Linear Control Theory

J. R. LEIGH

The Polytechnic of Central London

1980

ACADEMIC PRESS

A Subsidiary of Harcourt Brace Jovanovich, Publishers

London ∗ New York ∗ Toronto ∗ Sydney ∗ San Francisco

ACADEMIC PRESS INC. (LONDON) LTD.
24/28 Oval Road
London NW1

United States Edition published by
ACADEMIC PRESS INC.
111 Fifth Avenue
New York, New York 100003

British Library Cataloguing in Publication Data

Leigh, J. R.
Functional analysis and linear control theory.—
(Mathematics in science and engineering).
1. Control theory   2. Functional analysis
I. Title   II. Series
629.8′32′015157   QA402.3   80-41167

ISBN 0-12-441880-5

Printed in Great Britain by Page Bros (Norwich) Ltd., Norwich

Preface

The last twenty years have seen a rapid growth in control theory so that there now exists a formidable body of knowledge, particularly in the areas of multivariable control and optimal control. By the use of functional analysis, a concise conceptual framework can be put forward for linear control theory. Within this framework the unity of the subject becomes apparent and a wide range of powerful theorems becomes applicable. It is the aim of this book to assist mathematically interested graduate engineers to acquire this framework.

A difficulty for engineers in dealing with standard mathematical texts is the apparent lack of motivation in the development. I have tried to motivate this work by inserting remarks designed to help either motivation or visualization. Mathematicians wishing to obtain a concise overview of the structure of linear control theory will find the book well suited to this purpose. For such readers a start at Chapter 5 is recommended.

Chapter 1 is a rapid survey of basic mathematics. Chapters 2 to 5 contain most of the mathematics needed later in the book, although some theorems needed only in Chapter 10 are given in that chapter. This is because a course of study omitting Chapter 10 may sometimes be envisaged. Chapter 10 being the most demanding mathematically. Chapter 6 establishes axioms for linear dynamic systems and links the axiomatic description to the state space description. Stability, controllability and observability—important structural properties of a system—are considered in Chapter 7. Chapters 8 and 9 are concerned with the formulation of optimization problems and with the questions of existence, uniqueness and characterization of optimal controls. Chapter 10 is concerned with the extension to distributed parameter systems of some of the concepts and methods applied earlier to finite dimensional systems. A comprehensive set of references is given.

The approach and viewpoint of the book in relation to control theory can only be explained in function-analytic terms. The state space formulation of a linear system can be shown to be equivalent to a representation in terms of mappings between appropriate linear spaces (the input space, the state space, the output space), connected by two linear mappings (the transition mapping, the output mapping). In an optimization problem, the input space is shown to be a normed space, with the norm derived from the cost index of the problem. In this abstract setting the system reveals its underlying structure and the powerful methods of functional analysis can be applied directly.

Where the input space is a Hilbert space, the projection theorem allows useful results to be obtained quickly. Where the input space is a Banach space, the Hahn–Banach theorem is required and the dual space is frequently utilized to obtain inner-product-like quantities. Weak topologies are often required so that compactness arguments can be applied. Convexity, rotundity, smoothness and reflexivity are the properties of the spaces that appear in the existence and uniqueness proofs.

For the generalization to distributed parameter systems a wider range of tools is required including the theory of semi-groups and some background from spectral theory.

Acknowledgements

I am most grateful to Dr. M. R. Mehdi of Birkbeck College, University of London, for his interest and assistance in the development of this book. Not only did he check the whole of the manuscript, but he also made many suggestions that have helped the logical unfolding of the material in the early chapters.

Miss Susan Harding typed all the several versions of the manuscript and I am pleased to acknowledge my gratitude here.

The numerical example in Chapter 8 was computed by Mrs. S. A. P. Muvuti.

Contents

Preface   v

CHAPTER 1. **Preliminaries**   1

Control Theory. Set Theory. Linear Space (Vector Space). Linear
Independence. Maximum and Supremum. Metric and Norm.
Sequence and Limit Concepts. Convex Sets. Intervals on a Line. The
k Cube. Product Sets and Product Spaces. Direct Sum. Functions
and Mappings. Exercises.

CHAPTER 2. **Basic Concepts**   11

Topological Concepts. Compactness. Convergence. Measure Theory.
Euclidean Spaces. Sequence Spaces. The Lebesgue Integral. Spaces
of Lebesgue Integrable Functions (L^p Spaces). Inclusion Relations
between Sequence Spaces. Inclusion Relations between Function
Spaces on a Finite Interval. The Hierarchy of Spaces. Linear Functionals. The Dual Space. The Space of all Bounded Linear Mappings.
Exercises.

CHAPTER 3. **Inner Product Spaces and Some of their Properties**   26

Inner Product. Orthogonality. Hilbert Space. The Parallelogram Law.
Theorems. Exercises.

CHAPTER 4. **Some Major Theorems of Functional Analysis**   33

Introduction. The Hahn–Banach Theorem and its Geometric Equivalent. Other Theorems Related to Mappings. Hölder's Inequality.
Norms on Product Spaces. Exercises.

CHAPTER 5. **Linear Mappings and Reflexive Spaces**   41

Introduction. Mappings of Finite Rank. Mappings of Finite Rank on a Hilbert Space. Reflexive Spaces. Rotund Spaces. Smooth Spaces. Uniform Convexity. Convergence in Norm (Strong Convergence). Weak Convergence. Weak Compactness. Weak* Convergence and Weak* Compactness. Weak Topologies. Failure of Compactness in Infinite Dimensional Spaces. Convergence of Operators. Weak, Strong and Uniform Continuity. Exercises.

CHAPTER 6. **Axiomatic Representation of Systems**   50

Introduction. The Axioms. Relation between the Axiomatic Representation and the Representation as a Finite Set of Differential Equations. Visualization of the Concepts of this Chapter. System Realization. The Transition Matrix and some of its Properties. Calculation of the Transition Matrix for Time Invariant Systems. Exercises.

CHAPTER 7. **Stability, Controllability and Observability**   63

Introduction. Stability. Controllability and Observability. Exercises.

CHAPTER 8. **Minimum Norm Control**   79

Introduction. Minimum Norm Problems: Literature, Outline of the Approach. Minimum Norm Problem in Hilbert Space: Definition. Minimum Norm Problems in Banach Space. More General Optimization Problems. Minimum Norm Control: Characterization, a Simple Example. Development of Numerical Methods for the Calculation of Minimum Norm Controls. Exercises.

CHAPTER 9. **Minimum Time Control**   103

Preliminaries and Problem Description. The Attainable Set. Existence of a Minimum Time Control. Uniqueness. Characterization. The Pontryagin Maximum Principle. Time Optimal Control. Exercises.

CHAPTER 10. **Distributed Systems** 120

Introduction. Further Theorems from Functional Analysis. Axiomatic Description. Representation of Distributed Systems. Characterization of the Solution of the Equation $\dot{x} = Ax + Bu$. Stability. Controllability. Minimum Norm Control. Time Optimal Control. Optimal Control of a Distributed System: An Example. Approximate Numerical Solution. Exercises.

Glossary of Symbols 141

References and Further Reading 145

Subject Index 157

To Gisela

CHAPTER 1

Preliminaries

1.1 CONTROL THEORY

In linear theory we are in general interested in the behaviour of a system, here designated by the symbol Σ, and subject to the influence of one or more inputs, $u_1, \ldots, u_r$. The system is assumed to have n internal states $x_1, \ldots, x_n$ and m outputs $y_1, \ldots, y_m$ as illustrated in Figure 1.1.

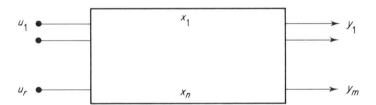

Figure 1.1.

In equation form, a model for this system is

$$
\begin{aligned}
\dot{x}(t) &= A(t)x(t) + B(t)u(t) \\
y(t) &= C(t)x(t)
\end{aligned}
\tag{1.1.1}
$$

with A, B, C being matrices that, as the notation indicates, can be time varying.

To obtain concrete results we shall use the above equation, but for more generalized study we shall usually use an axiomatic description of the system Σ and this is developed in Chapter 6.

Typical questions posed in control theory are:

(i) Can the system be steered from any initial state x_0 to any desired state x_d (controllability)?
(ii) Do there exist optimal control strategies for the steering of the system? If so, are they unique and how can they be synthesized?

(iii) For a particular control configuration, will the system Σ be stable?

(iv) Given a knowledge of the outputs y, can the values of the states x be inferred (observability)?

(v) Given a knowledge of the inputs u and the corresponding outputs y, can a mathematical model of the form (1.1.1) be produced to represent the system Σ (mathematical modelling)?

We shall concern ourselves with all these questions to some extent but question (ii) will be the most important.

The remainder of the chapter reviews the basic mathematical material required in the later chapters.

1.2 SET THEORY

$x \in E$ means that x is a member of the set E.

$x \notin E$ means that x is not a member of the set E.

The symbol $\varnothing$ is used to denote the empty set.

A set will often be defined in the following way

$$E = \{x | x \text{ has property } p\}$$

This is to be read as: E is the set of all x such that x has property p.

The set E is said to be a *subset* of the set F, written $E \subset F$, if every x belonging to E also belongs to F; in symbols

$$E \subset F \text{ if } x \in E \Rightarrow x \in F$$

Note that if $E \subset F$ and $F \subset E$, then necessarily $E = F$, a fact often used in proofs.

Let $E \subset F$, then the *complement* of the set E, denoted E' is defined

$$E' = \{x | x \in F, x \notin E\}$$

The *intersection* of a family of sets is denoted by $\bigcap$ and their *union* by $\bigcup$. Let $\{E_i\}$ be a family of sets then the definitions are as follows:

$$\bigcap E_i = \{x | x \text{ belongs to every } E_i\}$$

$$\bigcup E_i = \{x | x \text{ belongs to at least one of the sets } E_i\}$$

A *denumerable* (or countable) set is one that can be put into one to one correspondence with the set of positive integers. A *nondenumerable* (or uncountable) set has a higher *cardinality*, or "number" of elements, than a denumerable set.

1.3 LINEAR SPACE (VECTOR SPACE)

We denote by K the set of all scalars (real or complex).

Let Y be a set for which (i) addition of two elements y_1, y_2 is defined and for which $y_1 + y_2 \in Y$; (ii) multiplication of any element $y \in Y$ by a scalar $\alpha \in K$ is defined and $\alpha y \in Y$.

The operations of addition and scalar multiplication are assumed to satisfy the usual fundamental axioms. Then Y is called a *linear space* (vector space) over K. Let Z be a subset of y such that Z is a linear space in its own right then Z is called a *linear subspace* of Y.

1.4 LINEAR INDEPENDENCE

Let Y be a linear space over K and let $y_1, \ldots, y_n$ be elements in Y. The set $\{y_1, \ldots, y_n\}$ is said to be *linearly independent* if the equation $\sum_{i=1}^{n} \alpha_i y_i = 0$ implies that $\alpha_i = 0$, $i = 1, \ldots, n$. (Geometrically we can imagine that no linear combination of the vectors y_i can lead us to the origin for any choice of the α_i except when all the $\alpha_i = 0$).

If the set is not linearly independent then it is said to be *linearly dependent*. If there are n, and no more than n, linearly independent elements in Y, then we say that Y has dimension n or is an n dimensional linear space. Thus, the *dimension* of a linear space Y is the maximum number, n, of linearly independent elements $(y_1, \ldots, y_n)$ in Y. Such a set of linearly independent elements is called a *basis* for Y. The n dimensional space of real n vectors is *denoted by* $\mathbb{R}^n$. Further information on the $\mathbb{R}^n$ spaces is given in Section 2.5.

1.5 MAXIMUM AND SUPREMUM

A *sequence* is an indexed set of elements α_1, $\alpha_2, \ldots$, denoted $\{\alpha_i\}$, it being understood that i runs through the positive integers.

Let $\{y_i\}$, $i = 1, \ldots, n$, be a finite sequence of real numbers. We define

$$\max_i \{y_i\} = y_j \quad \text{provided that } y_j \geqslant y_i, \quad i = 1, \ldots, n$$

Clearly at least one such element must exist. Now let $\{y_i\}$, $i = 1 \ldots$, be an infinite sequence of real numbers. To write $\max_i \{y_i\}$ may not be meaningful since, e.g., if $y_i = 1 - (1/i)$, then for i large, y_i is very close to unity but does not attain that value. For infinite sequences we define the *supremum* (sup) as follows. Given the sequence $\{y_i\}, i = 1, \ldots,$ let m be any number satisfying $m \geqslant y_i, \forall i$ then m is called an *upper bound* for the sequence $\{y_i\}$. Now define m_0 to be the smallest of the upper bounds; then m_0 is called the *supremum* of $\{y_i\}$. (Also called the *least upper bound*.) The supremum of a real function f on an interval $[a, b]$ is defined in a similar way. We denote by $\sup_x \{f(x)\}$ the least value of m satisfying $m \geqslant f(x), \forall x \in [a, b]$. A function may or may not attain its supremum—an important point that is directly related to the existence of well defined optimal control strategies. The *infimum* or *greatest lower bound* of a function f is defined analogously as the greatest value of m satisfying $m \leqslant f(x), \forall x \in [a, b]$.

1.6 METRIC AND NORM

The distance between two elements of an arbitrary set is not defined. However, if a metric, defined below, is introduced to the set, a property analogous to distance then exists. Let y_1, y_2 be any elements of a set Y then the real valued function $d(y_1, y_2)$ defined on Y is a *metric* if it satisfies the conditions

$$d(y_1, y_2) = 0, \quad y_1 = y_2; \qquad d(y_1, y_2) > 0, \quad y_1 \neq y_2$$

$$d(y_1, y_2) = d(y_2, y_1)$$

$$d(y_1, y_2) \leqslant d(y_1, y_3) + d(y_3, y_2)$$

The pair $\{Y, d\}$ is called a *metric space*.

Let $f(t), g(t)$ be two integrable functions of time on an interval $[a, b]$ then $d(f, g)$ can be defined with equal validity by, for example,

$$d(f, g) = \sup_{t \in [a, b]} |f(t) - g(t)|$$

$$d(f, g) = \int_a^b |f(t) - g(t)| \, dt$$

$$d(f, g) = \left(\int_a^b |f(t) - g(t)|^2 \, dt \right)^{1/2}$$

According to the physical background of the problem, a particular metric can be selected to be the most meaningful.

The *norm* of an element y in a linear space Y, denoted $\| y \|$, is a measure of the magnitude of the element and is a generalization of the concept of modulus or absolute value.

The axioms for a norm are

(i) $\| y \| = 0$ if and only if $y = 0$ (i.e. if y is the zero vector);

(ii) $\| \alpha x \| = |\alpha| \| x \|$, ($\alpha \in K, x \in Y$);

(iii) $\| x + y \| \leqslant \| x \| + \| y \|$, $\forall x, y \in Y$.

A linear space on which a norm has been defined is called a *normed space*.

1.7 SEQUENCES AND LIMIT CONCEPTS

A sequence $\{y_i\}$ in a metric space Y is said to be *convergent* to an element y if given $\varepsilon > 0$ there exists an integer N such that $d(y_i, y) < \varepsilon$ whenever $i > N$. When dealing with sequences, two important questions arise. Is the sequence convergent and, if so, is the limit of the sequence y a member of the space Y. To proceed further, we need the concept of a Cauchy sequence.

Let Y be a metric space with metric d. Let $\{y_i\}$ be a sequence each of whose elements is in Y and for which, given $\varepsilon > 0$, there exists an integer N such that $m, n > N \Rightarrow d(y_m, y_n) < \varepsilon$. Then $\{y_i\}$ is called a *Cauchy sequence*. It is easy to see that every convergent sequence is a Cauchy sequence. However, the converse is not always true but we have the important result. The space Y is called a *complete space* if whenever the sequence $\{y_i\}$ is a Cauchy sequence then $\{y_i\}$ converges to an element y and y is a member of the space Y. This statement defines completeness, which is property of such importance that spaces not possessing the completeness property will be met only rarely in this book.

Let Y be a metric space and E be a subset of Y. The set E is said to be *bounded* if $\exists a \in Y$ such that $d(x, a) < \infty, \forall x \in E$.

Let X be a metric space and x be a point in X. A subset $N_\varepsilon(x)$ of X defined

$$N_\varepsilon(x) = \{y | d(y, x) < \varepsilon\}$$

is called a *neighbourhood of x*. (Note that alternative and more generalized definitions of neighbourhood will be encountered in other texts.)

Let $\{x_i\}$ be a sequence of points in a metric space X; then the sequence converges to the point x if for every neighbourhood $N_\varepsilon(x)$ of x, there exists an integer M, such that $x_i \in N_\varepsilon(x), \forall i > M$. The sequence $\{x_i\}$ is convergent to x and x is called the *limit of the sequence*.

Let E be a subset of a metric space X then a point $x \in E$ is called an *interior point* of E if for some $\varepsilon > 0$ there exists a neighbourhood $N_\varepsilon(x) \subset E$. The set of all interior points of E is called the *interior of E*, denoted E^0. If $E = E^0$ then the set E is defined to be *open*.

Let X be a metric space and E be a subset of X. The point $x \in X$ is called a *boundary point* of E if every neighbourhood of x contains points of E and points of E'. The set of all boundary points of E is called the *boundary* of E, denoted ∂E.

Let A be any set in a metric space Y and let z be a point which may or may not belong to A. Suppose that every neighbourhood of z contains infinitely many points of A, then z is called a *limit point* of A. Intuitively, let $\{y_i\}$ be a convergent sequence in A, then $\lim_{i \to \infty} \{y_i\}$ would necessarily be a limit point of the set A. We note the following: A set is *closed* if it contains all its own limit points. The *closure* of the set Y, denoted $\bar{Y}$, is the set Y plus all the limit points of Y. A set Y is *dense* in a set Z if $\bar{Y} = Z$. In such a case, elements in Z can be approximated arbitrarily well by elements in Y (cf. the approximation of real numbers to any accuracy by rationals—the rational numbers being dense in the reals).

1.8 CONVEX SETS

Convex sets are particularly important in optimal control theory. The set E is *convex* if given any two points $x_1, x_2 \in E$ and any real number α with $0 \leqslant \alpha \leqslant 1$ we have

$$\alpha x_1 + (1 - \alpha)x_2 \in E.$$

(Geometrically the line joining any two points in a convex set E lies wholly in E.)

1.9 INTERVALS ON A LINE

Given two real numbers $a, b; a < b$, define the sets

$$E_1 = \{\text{all real numbers } x \,|\, a \leqslant x \leqslant b\}$$

$$E_2 = \{\text{all real numbers } x \,|\, a < x < b\}$$

$$E_3 = \{\text{all real numbers } x \,|\, a \leqslant x < b\}$$

$$E_4 = \{\text{all real numbers } x \,|\, a < x \leqslant b\}$$

$E_1, \ldots, E_4$ are called intervals. E_1 is called a *closed interval*, E_2 is called an *open interval*, E_3 and E_4 are half open intervals.

Notation

$$E_1 \text{ is denoted } [a, b]$$

$$E_2 \text{ is denoted } (a, b)$$

$$E_3 \text{ is denoted } [a, b)$$

$$E_4 \text{ is denoted } (a, b]$$

1.10 The k CUBE

Let U be an r dimensional space. The k cube in U is defined to be the set

$$\left\{ u \middle| u = \begin{pmatrix} u_{,1} \\ \vdots \\ u_r \end{pmatrix}, \quad |u_i| \leqslant k, \quad i = 1, \ldots, r, \quad u \in U \right\}$$

1.11 PRODUCT SETS AND PRODUCT SPACES

Let X, Y be any two sets. Then the *product set* denoted $X \times Y$ is the set of all ordered pairs (x, y), $x \in X$, $y \in Y$.

If X and Y are both metric spaces then a metric for the product set can be formed in various ways by combining the separate metrics of X and Y. Then $X \times Y$ is called the *product metric space.*

1.12 DIRECT SUM

Let X, Y be two linear spaces then a linear space called the *direct sum* of X and Y and denoted $X \oplus Y$ is formed by taking the product $X \times Y$ of all ordered pairs and defining addition and scalar multiplication on this set. If two elements of $X \oplus Y$ are (x_1, y_1) and (x_2, y_2) then addition of these elements would result in the new element $(x_1 + x_2, \ y_1 + y_2) \in X \oplus Y$. Scalar multiplication is defined in the obvious way:

$$\alpha(x, y) = (\alpha x, \alpha y), \quad \alpha \in K$$

1.13 FUNCTIONS AND MAPPINGS

Given two sets Y, Z a *function f* can be considered to be a rule that, given some element $y \in Y$, assigns a unique element $z \in Z$. We write $z = f(y)$. The set of $y \in Y$ for which the function assigns an element of Z is called the *domain* of f, denoted $D(f)$. The set $\{z | z = f(y), y \in D(f)\}$ is called the *range* of f. In case Range $(f) = Z$ the function f is called a *surjective function*. If, for every element $z \in$ Range (f), there exists a unique element y such that $z = f(y)$ then f is said to be an *injective function*. If f is both surjective and injective then f is said to be a *bijective function*. If two sets Y, Z are linked by a bijective function, then the two sets Y, Z are said to be in *one to one corres-pondence*. The *inverse* of an injective function f is denoted f^{-1} and is defined by the relation $f^{-1}(f(y)) = y, f(f^{-1}(z)) = z$.

A function on a linear space to another linear space is defined to be *linear* if it satisfies the two relations:

(i) $f(\alpha y) = \alpha f(y)$, $\forall y \in Y$, α a real number;

(ii) $f(y_1 + y_2) = f(y_1) + f(y_2)$, $\forall y_1, y_2 \in Y$.

We define a function f to be *locally linear* if for some positive constant M.

(i) $f(x_1 + x_2) = f(x_1) + f(x_2)$, $\forall x_1, x_2 \in X$ such that

$$\|x_1\| + \|x_2\| \leqslant M$$

(ii) $f(\lambda x) = \lambda f(x)$, $\forall x \in X$, $\forall \lambda \in \mathbb{R}^1$ such that

$$\|x\| \leqslant M, \|\lambda x\| \leqslant M$$

This definition is useful since all practical systems saturate in the presence of a sufficiently large input signal. A locally linear system can be treated as a linear system provided that the saturation constraint is not violated.

Let $z = f(y)$ then we say that z is the *image* of y under f. Conversely y is said to be the pre-image of z.

Let X, Z be metric spaces. The function $f: X \rightarrow Z$ is *continuous* if for all $y \in D(f)$, given $\varepsilon > 0$, $\exists \delta(y) > 0$ such that $d(f(x), f(y)) < \varepsilon$ for all x satisfying $d(x, y) < \delta(y)$; δ is a function of y as the notation indicates, but if for every $y \in D(f)$, given $\varepsilon > 0$ there exists a constant $\delta > 0$ such that $d(f(x), f(y)) < \varepsilon$ for all x satisfying $d(x, y) < \delta$, then f is said to be *uniformly continuous*.

Notice that $d(x, y)$, $d(f(x), f(y))$ are respectively the metrics for X and Z.

A necessary and sufficient condition that a function f between metric spaces be continuous at a point x is that $\{x_n\} \rightarrow x \Rightarrow f(x_n) \rightarrow y = f(x)$. In words, that the image under f of the elements of a convergent sequence is also a convergent sequence and that the image of the limit is equal to the limit of the sequence of images.

Let Y, Z be linear spaces over the same scalar field then a *linear mapping* T is simply a linear function $T: Y \rightarrow Z$.

Very often if the linear mapping T maps Y into itself, $T: Y \rightarrow Y$ then T is called a *linear operator*.

Let Y, Z be normed spaces. The linear mapping $T: Y \rightarrow Z$ is *bounded* if for every $y \in Y$, $\|T(y)\|/\|y\| \leqslant m < \infty$. Every bounded linear mapping is a *continuous linear mapping* and conversely. By tradition these mappings are referred to as bounded rather than continuous.

The *operator norm* of a linear mapping, denoted $\|T\|$ is defined by

$$\|T\| = \sup_{y \in Y} \frac{\|T(y)\|}{\|y\|}$$

The *null space* of a linear mapping T is denoted N_T and is defined by

$$N_T = \{y \,|\, T(y) = 0\}$$

1.14 EXERCISES

(1) Let

$$E_1 = \{1, 2, 3, 4\}, \qquad E_2 = \{1, 3, 5\}, \qquad F = \{1, 2, 3, 4, 5, 6\}$$

Consider E_1, E_2 as subsets of F.
  (a) Which elements are in the set $(E_1 \cap E_2)'$?
  (b) Which elements are in the set $(E_1 \cup E_2)'$?

(2) Show that the set of vectors

$$\begin{bmatrix} x_1 \\ x_2 \end{bmatrix}$$

in the plane is a linear space over K. Give an example of a linear subspace in the plane.

(3) Prove that any three vectors in $\mathbb{R}^2$ must be linearly dependent. Sketch the geometric interpretation of the test for linear dependence.

(4) Consider the power series $x + x^2 + x^3 + \ldots + x^n$ with $x \in [0, 1]$. Are the terms linearly independent?

(5) Prove that the set $\{1/i\}$ $i = 1, \ldots,$ has no minimum value. What is the infimum?

(6) Let X be the set of all ordered pairs of real numbers. Consider how three metrics, similar to d_1, d_2, d_3 in the text, can be realised in X.

(7) Let $X = \mathbb{R}^n$ and let $x, y \in X$. Let d_1, d_2, d_3 be the three metrics defined in Section 1.6. Show that $d_3(x, y) \leqslant d_2(x, y) \leqslant d_1(x, y)$.

(8) Define a function $f : \mathbb{R}^1 \to \mathbb{R}^1$ by $f(x) = |x|$. Show that f is continuous.

(9) Let E be the interval $(0, 1)$. Find a Cauchy sequence in E that has no limit in E.

(10) Prove that if $\{x_n\}$ is a convergent sequence in a metric space X then $\{x_n\}$ is a Cauchy sequence.

(11) Let $X = \mathbb{R}^1$. Each of the functions f_i defined below is a real valued function $f_i : X \to X$,

$$f_1(x) = 1/x, \quad x \neq 0; \quad f_1(0) = 1 \qquad f_2(x) = k, \quad k \text{ a constant}$$

$$f_3(x) = \cos x \qquad f_4(x) = \tan x$$

$$f_5(x) = \sin(1/x), \quad x \neq 0 \qquad f_5(0) = 1$$

$$f_6(x) = x^{1/2}, \quad x \geq 0 \qquad f_6(x) = 0, \quad x < 0$$

$$f_7(x) = 1, x > 0 \qquad f_7(x) = -1, \quad x < 0 \qquad f_7(x) = 0, \quad x = 0$$

$$f_8(x) = x, \quad x > 0 \qquad f_8(x) = -x, x < 0 \qquad f_8(x) = 0, \quad x = 0$$

$$f_9(x) = x^2 \qquad f_{10}(x) = kx$$

with $\mathbb{R}^1$ as the domain, determine which of the functions is: (a) linear, (b) continuous, (c) uniformly continuous, (d) differentiable, (e) injective, (f) surjective. Which of the continuous functions map open sets into open sets?

If the domain is restricted to $[-1, 1]$ which of the functions now become uniformly continuous?

(12) Let $T : X \to Y$ be a linear mapping where X and Y have finite dimension. Prove that $\dim X = \dim Y + \dim N_T$, where N_T is the nullspace of T.

(13) Let X, Y be linear spaces and let $T : X \to Y$ be a linear mapping from X onto Y. Show that T is a bijection if and only if $N_T = \{0\}$.

CHAPTER 2

Basic Concepts

This chapter concentrates on the main concepts necessary to a proper understanding of later chapters. In reading this chapter, the emphasis should be on visualization and it will often be useful to attempt an informal sketch illustrating each new property encountered.

2.1 TOPOLOGICAL CONCEPTS

The important concepts of continuity, convergence and compactness are topological concepts. Topological properties are invariant under smooth mappings and as such they can be considered without reference to any metric.

Let Y be any set. We define axiomatically certain subsets of Y, called by convention the *open sets* of Y. The collection of open sets is called a *topology* for Y and is usually denoted by τ. We call τ a topology if it satisfies the following conditions

(i) $Y \in \tau, \varnothing \in \tau$.
(ii) Any union of sets in τ belongs to τ.
(iii) The intersection of a *finite* number of sets in τ belongs to τ.

The pair (Y, τ) is called a *topological space*.

In case Y is already a linear space, (Y, τ) is then called a *topological vector space*, provided that addition and scalar multiplication are continuous.

A set E in Y is defined to be a *closed set* if its complement E' is an open set. We see that the whole space Y and the empty set $\varnothing$ are both open sets and closed sets. Thus, particular sets can be both open and closed while in general an arbitrary set in Y need not be either open or closed.

At one extreme we can define Y and $\varnothing$ to be the only open sets and at the other extreme we can define every possible set in Y to be open. If two topologies τ_1, τ_2 are defined on the same space Y and if every open set in τ_1 is also a member of τ_2 but not conversely, then τ_2 has more open sets than does τ_1.

11

τ_2 is a *stronger topology* than τ_1. τ_1 is a *weaker topology* than τ_2. The significance of the foregoing from our point of view is that a set may fail to be compact in one topology but be compact in a weaker topology. Working within the framework of the weaker topology, theorems requiring compactness can still be applied.

Topology was developed as an axiomatization and generalization of easily understood geometric concepts. Let Y be the Euclidean plane with the usual distance measurement between two points (x_1, y_1), (x_2, y_2) being defined by

$$d((x_1, y_1), (x_2, y_2)) = ((x_1 - x_2)^2 + (y_1 - y_2)^2)^{1/2}$$

The usual way of defining the collection τ of open sets in Y is then in terms of this metric,

$$\tau = \{y | d(y, z) < r, \qquad y, z \in Y, \quad r > 0\}$$

Thus, the open sets for Y are disks, centred at some point z in Y and of radius r with the boundary of the disk being specifically excluded. It is the exclusion of the boundary that is intuitively the best characterization of openness. Conversely, it is the inclusion of the boundary that characterizes closedness. Because we shall use frequently the concept of a *closed ball* of radius r centred at a point z we use the special notation $\beta_r(z)$ to indicate such a closed ball. It is defined fairly obviously by the relation

$$\beta_r(z) = \{y | d(y, z) \leqslant r, y \in Y\}$$

(a ball of unit radius centred at the origin is denoted by β).

There are "weak" analogues of several of the well known theorems of analysis—the usefulness of the weak analogues is that they are often applicable when the "strong" theorems are inapplicable. For instance, in the weak topology a set is more likely to be compact and a function more likely to be continuous, than in the norm topology.

2.2 COMPACTNESS

Compactness is a topological property that is of fundamental importance, yet it is difficult to form an intuitive, geometric picture of this property.

The basic definition of compactness is as follows.

Let E be a set in a metric space Y (or more generally in a topological space). A collection of open sets $\{M_i\}$, $i \in \mathbb{R}^1$ is said to be an open covering for E if $E \subset \cup_i M_i$.

The set E is defined to be *compact* if every open covering for E contains a finite subcovering. In symbols, E is compact if for every open covering $\{M_i\}$ for E there exists an integer n such that $E \subset \bigcup_{i=0}^{n} {}^1 M_i$.

Let E be a set in $\mathbb{R}^n$ then E is compact if and only if E is closed and bounded. However, in general, compactness is a very strong requirement and for instance, for a function space, in the natural topology generated by its own metric, a closed bounded set will not be compact.

Let E be a compact set in a metric space Y. Then every sequence $\{e_i\}$ in E, contains at least one convergent subsequence, convergent to a point in E. (A *subsequence* is simply an infinite selection from the original sequence with the order being maintained.) In a metric space, this property characterizes compactness.

In the case of a real sequence we can visualize that $\{e_i\}$, if not convergent, must eventually oscillate, since if a sequence is not convergent it must go to $+\infty$, $-\infty$, or must oscillate and the first two possibilities are eliminated by the compactness of E.

Preservation of compactness under bounded linear mappings

Let E be a compact set in a metric space Y. Let T be a continuous mapping $T: Y \to Z$ to another metric space Z. Define the set

$$F = \{z | z \in Z, z = Te, e \in E\}$$

then the set F is compact.

2.3 CONVERGENCE

2.3.1 Pointwise convergence

Let $\{f_i\}$ be a sequence of functions on a metric space X taking values in a metric space Y. If there exists a function $f: X \to Y$ such that for every $x \in X$, $d(f_i(x), f(x)) \to 0$ as $i \to \infty$ then $\{f_i\}$ is said to *converge pointwise* to the function f.

It is important to note that the limit function f of a sequence may be discontinuous even though each member of the sequence $\{f_i\}$ is continuous. The classic example of this is $f_i(x) = x^i$ defined on $[0, 1]$ with the metric

$$d(f_i, f_j) = |f_i(x) - f_j(x)|$$

which has a discontinuous function f as its (pointwise) limit, where

$$f(x) = 0, \qquad 0 \leqslant x < 1$$

$$f(1) = 1$$

Pointwise convergence is not a strong enough property to ensure that continuity is carried over from the functions f_i to the pointwise limit f.

2.3.2 Uniform convergence

Again, let $\{f_i\}$ be a sequence of functions on a metric space X taking values in a metric space Y. If there exists a function $f : X \to Y$ such that

$$\sup_{x \in X} (d(f_i(x), f(x))) \to 0 \quad \text{as } i \to \infty$$

then the sequence $\{f_i\}$ is said to *converge uniformly* to the function f.

It is easy to see that the sequence of functions $\{x^i\}$ defined on $[0, 1]$ above is not uniformly convergent to the function f.

In fact,

$$\sup_{\substack{x \\ i \to \infty}} |f_i(x) - f(x)| = \lim_{\substack{x \to 1 \\ i \to \infty}} |f_i(x) - f(x)| = 1$$

and the sequence $\{f_i\}$ is not uniformly convergent.

Under uniform convergence, the following properties hold for sequences of functions.

(1) If a sequence of functions $\{f_i\}$ is uniformly convergent to a function f and if each function f_i is continuous, then the limit function f is continuous. (Note, however, that uniform convergence is sufficient but not necessary to ensure continuity of f.)

(2) Let $\{f_i\}$ be a sequence of real valued integrable functions defined on an interval $[a, b]$. Let the sequence be uniformly convergent to a limit function f then

$$\lim_{i \to \infty} \int_a^b f_i \, dt = \int_a^b f \, dt, \qquad t \in [a, b]$$

(3) Let $\sum_{i=1}^{\infty} e_i$ be a uniformly convergent series with each e_i being integrable over $[a, b]$ then

$$\int_a^b \sum_{i=1}^{\infty} e_i \, dt = \sum_{i=1}^{\infty} \int_a^b e_i \, dt, \qquad t \in [a, b]$$

Uniform convergence of the series $\sum_i e_i$ is sufficient (but not necessary) to ensure the validity of term by term integration.

(4) Let $\{f_i\}$ be a sequence of real valued functions defined on a closed interval $[a, b]$, convergent pointwise to a function f and differentiable there. If $\{f_i'\}$ converges uniformly on $[a, b]$ then f is differentiable on $[a, b]$ and $\lim_{i \to \infty} \{f_i'\} = f'$ (where $'$ denotes derivative).

In summary, we can say that under conditions of uniform convergence:

(i) Continuity in members of a convergent sequence carries over to the limit function;
(ii) It is permissible to interchange the order of two limit operations.

2.4 MEASURE THEORY

Measure theory is concerned with the problem of assigning a "size" or "content" to point sets. Let E be a point set then the *measure* of E, denoted $\mu(E)$ is a non-negative number measuring its size or volume. Measure theory is particularly important as a foundation for probability theory and for the development of integration theories. In this book, or in similar literature, measure theory plays a relatively minor role.

Different measures have been defined, applicable to abstract sets. The most important measure is that of Lebesgue, and here we define *Lebesgue measure on the real line*.

Let (a, b) be an open interval in $\mathbb{R}^1$ then the measure of (a, b) denoted $\mu(a, b) = b - a$. Let E be a non-empty bounded open set in $\mathbb{R}^1$ then

$$\mu(E) = \sum_i \mu(\delta_i)$$

where δ_i are the disjoint component intervals in E (the set $\{\delta_i\}$ may be finite or denumerable). The measure of a closed set F then follows by containing F in an open set E so that $F \subset E$ and using the relation

$$\mu(F) = \mu(E) - \mu(E - F)$$

both sets on the right-hand side of the equation being open.

The *outer measure* of a bounded set E in $\mathbb{R}^1$ is denoted $\mu^*(E)$ and is defined by

$$\mu^*(E) = \inf_{G \supset E} \{\mu(G)\}.$$

where $\{G\}$ is the collection of all bounded open sets for which $E \subset G$. Similarly, the *inner measure* $\mu_*(E)$ of a bounded set E in $\mathbb{R}^1$ is defined by

$$\mu_*(E) = \sup_{F \subset E} \{\mu(F)\}$$

where $\{F\}$ is the collection of all closed sets for which $E \subset F$.

A bounded set E in $\mathbb{R}^1$ is said to be *measurable* if $\mu^*(E) = \mu_*(E)$ and the measure of E is the common value of outer and inner measure, denoted $\mu(E)$. It can be shown that when E is open or closed this definition agrees with the earlier one given above.

We define a *null set* as one that has measure zero. In other words, A is a null set if $\mu(A) = 0$. Now let E be a set of points, each of which may or may not possess a property p. Then if the set of points in E not possessing the property is a null set, we say that the property p holds *almost everywhere* on E.

2.5 EUCLIDEAN SPACES

Let $\mathbb{R}^1 = \{$the set of all real numbers$\}$. $\mathbb{R}^2$ is the product space $\mathbb{R}^1 \times \mathbb{R}^1$ and is the set of all ordered pairs (x, y) $x \in X = \mathbb{R}^1$, $y \in Y = \mathbb{R}^1$. Similarly, $\mathbb{R}^n = \mathbb{R}^1 \times \mathbb{R}^1 \times \dots \mathbb{R}^1$, n times, and is the set of all n-tuplets of real numbers. The norm in $\mathbb{R}^1$ is the absolute value, i.e.

$$\|x\| = |x|$$

Let

$$x = \begin{bmatrix} x_1 \\ \vdots \\ x_n \end{bmatrix} \in \mathbb{R}^n$$

then

$$\|x\| = (x_1^2 + x_2^2 + \dots + x_n^2)^{1/2}.$$

2.6 SEQUENCE SPACES

Let X be the set of all sequences $\{x_i\}$ of real numbers. Let every element in X satisfy

$$\left(\sum_{i=1}^{\infty} |x_i|^p \right)^{1/p} = M < \infty$$

with p a real number satisfying $1 \leqslant p < \infty$. Then M is a norm for X and X is called an l^p *space*. When $p = \infty$ we define $\|x\| = \sup_i \{|x_i|\}$.

2.7 THE LEBESGUE INTEGRAL

Consider the function f defined by

$$f(x) = k \text{ when } x \in [0, 1]$$

$$x \in [2, 3] \quad \text{(where } k \text{ is a constant)}$$

$$f(x) = 0 \qquad \text{otherwise.}$$

Suppose we wish to evaluate

$$\int_0^4 f(x)\,dx.$$

Clearly the numerical value of the integral is $2k$. We used the rule, numerical value of integral = (total length of interval on which $f(x)$ takes on its non-zero value) $\times k$. Next consider the function $g(x)$ defined by

$$g(x) = 0, \qquad x \text{ irrational}$$

$$g(x) = 1, \qquad x \text{ rational}$$

and attempt to evaluate the integral

$$\int_0^1 g(x)\,dx$$

We might expect by analogy with the first example that $\int_0^1 g(x)\,dx =$ ("length" of interval within $[0, 1]$ on which x is rational) $\times 1$, which we cannot evaluate by elementary means.

In fact, the Riemann integral, which is the one used in elementary mathematics, defined as the limit of the area of a sum of rectangles, does not exist for $g(x)$.

However, it is legitimate to write:

$$\int_0^1 g(x)\,dx = (\text{measure within } [0, 1] \text{ of the set on which } x \text{ is rational}) \times 1.$$

Now measure is a formalization and generalization of the concept of the length of an interval along the real line, and provided that the statement in the bracket is meaningful, i.e. that the measure of the set exists, then so will the integral exist. (It is clear, reconsidering the first example above, that the measure of a simple interval must coincide with its length if contradictions are to be avoided.)

The integral as usually defined and used in elementary mathematics is the Riemann integral. It is quite possible to define the space of all Riemann integrable functions but a difficulty arises in that there exist convergent sequences of Riemann integrable functions whose limit is not Riemann integrable. In other words, the space of Riemann integrable functions is not complete. Since many proofs require completeness, the space of Riemann integrable functions is not in general a useful concept. The *Lebesgue integral* overcomes these problems. It has the following properties:

(i) The space of Lebesgue integrable functions is complete.
(ii) Every function that is Riemann integrable is Lebesgue integrable and the two integrals are numerically equal.

From the point of view of this book, little more needs to be known about the Lebesgue integral but it would be natural for the reader to wish to know more and a good reference would be M6.

2.8 SPACES OF LEBESGUE INTEGRABLE FUNCTIONS (L^p SPACES)

Let p be a number in the interval $[1, \infty)$ then the space of functions for which

$$\left(\int_a^b |f(t)|^p \, dt \right)^{1/p}$$

exists is denoted by $L^p[a, b]$.

The L^p spaces are linear (vector spaces) since the sum of two integrable functions is again integrable and the integral of a scalar multiple of a function is the scalar multiple of the integral of the function. In symbols, given f_1, f_2 for which

$$\left(\int_a^b |f_1(t)|^p \, dt \right)^{1/p} < \infty, \qquad \left(\int_a^b |f_2(t)|^p \, dt \right)^{1/p} < \infty$$

we have

$$\left(\int_a^b |f_1(t) + f_2(t)|^p \right)^{1/p} < \infty$$

Given a real number α we have

$$\left(\int_a^b |\alpha f_1(t)|^p \, dt \right)^{1/p} = |\alpha| \left(\int_a^b |f_1(t)|^p \, dt \right)^{1/p}$$

The L^p spaces are metric spaces with $d(f_1, f_2)$ defined by

$$d(f_1, f_2) = \left(\int_a^b |f_1(t) - f_2(t)|^p \, dt \right)^{1/p}$$

The L^p spaces are normed spaces with $\|f\|$ defined by

$$\|f\| = \left(\int_a^b |f(t)|^p \, dt \right)^{1/p}$$

However, note that in L^p spaces we are always dealing with equivalence classes of functions rather than with individual functions. The reasons for this are discussed below.

Let f_1, f_2 be two functions in an L^p space then as noted before

$$d(f_1, f_2) = \left(\int_a^b |f_1(t) - f_2(t)|^p \, dt \right)^{1/p}$$

We also know from the axioms for a metric that $d(f_1, f_2) = 0$ if and only if $f_1 = f_2$.

Now suppose, for example, that $f_1(t) = t$ on $[0, 2]$ while $f_2(t) = t$ on $[0, 1), (1, 2]$ with $f_2(1) = 0$, then, although clearly the two functions are not identical, we obtain $d(f_1, f_2) = 0$ which is a contradiction. We can see that functions differing only at isolated points cannot be distinguished in an L^p space. More formally, functions differing on a set of measure zero have the same Lebesgue integral and cannot be distinguished.

Thus, when we speak of a function f in an L^p space we are really talking about a set of functions $\{g | d(g, f) = 0\}$. This set of functions is an equivalence class since let $\circ$ be the relation "belongs to the same set as" then clearly

$$f \circ f \text{ (reflexive property)}$$

$$f \circ g \Rightarrow g \circ f \text{ (symmetric property)}$$

$$f \circ g, g \circ h \Rightarrow f \circ h \text{ (transitive property)}$$

these three properties ensuring that the set is an equivalence class.

We also note in passing that if two functions f_1, f_2 are members of the same L^p space then although we may have $d(f_1, f_2) = 0$ it is still possible that

$$\sup_{t \in (a, b)} f_1(t) \neq \sup_{t \in (a, b)} f_2(t)$$

Again, this arises because the functions differ at isolated points. To overcome this problem, the *essential supremum* (ess sup), is defined by disregarding isolated points

$$\text{ess sup}_t = \sup_t \{ f(t) | t \in (a, b), \text{ isolated points of } f(t) \text{ being disregarded} \}.$$

The L^p spaces are complete in the norm and it is probably sufficient for the purposes of this book to say that the L^p spaces are defined so as to be complete. The proofs of the other properties above are easy. It should be noted that if $p < 1$, the triangle inequality fails and both the norm and the metric break down.

A complete normed space is called a *Banach space* and the L^p spaces, $1 \leqslant p \leqslant \infty$, are the most important Banach spaces both from the point of view of this book and in general.

2.9 INCLUSION RELATIONS AMONGST SEQUENCE SPACES

Let $\mathscr{C}$ be the set of all convergent sequences. Let $\mathscr{C}_0$ be the set of all sequences convergent to zero. Then we have the following inclusion relations amongst the sequence spaces

$$l^1 \subset l^p \subset l^q \subset \mathscr{C}_0 \subset \mathscr{C} \subset l^\infty$$

where $1 \leqslant p \leqslant q$. Clearly, every convergent sequence is bounded and every summable sequence is convergent to zero so it is only the first three relations that are not obvious by inspection. The reader is asked to verify these relations.

2.10 INCLUSION RELATIONS AMONGST FUNCTION SPACES ON A FINITE INTERVAL

Let P be the space of all polynomials. Let C^n be the space of all n times differentiable functions. Let C be the space of all continuous functions. Let $1 < p < q < \infty$, then we have the inclusion relations, assuming that all the functions are defined on the same finite interval.

$$P \subset C^\infty \subset C^1 \subset C \subset L^\infty \subset L^q \subset L^p \subset L^1$$

These relations are fairly obvious. For instance, $P \subset C^\infty$ means that every polynomial is infinitely differentiable. $C \subset L^\infty$ means that every continuous function is bounded.

The relations between the L^p spaces are more interesting. The reader should satisfy himself of the validity of these last relations.

2.11 THE HIERARCHY OF SPACES

The spaces that are usually encountered in control theory have a considerable structure and it is helpful to understanding to visualize the nature of the structure.

(1) A set on which addition of elements and multiplication of an element by a scalar has been properly defined is called a linear space or vector space. Notice that this is a purely algebraic concept.

(2) A set on which a collection of open sets has been properly defined is called a topological space. It is often revealing to work in a topological space, where distance is not defined and coordinates do not exist, since one is led to the most fundamental formulation. A surprising number of results are still possible in this very general setting.

(3) A set on which distance between elements has been properly defined is called a metric space. Definitions of distance very different from that usually used are perfectly acceptable mathematically. For instance, let x_1, x_2 be two elements in X. An acceptable definition of distance (metric)

for X is that

$$d(x_i, x_j) = 1, \qquad i \neq j$$
$$d(x_i, x_j) = 0, \qquad i = j$$

where d is the distance between the two elements. A normed space is also a metric space with $d(x, y) = \|x - y\|$.

(4) A set on which the properties (1) and (2) have been defined in a compatible way is called a *topological vector space*.

(5) Every metric space X has a *natural topology* formed by constructing open sets of the form $\{x | d(x, x_0) < \varepsilon\}$ at any point x_0. The natural topology for a metric space is also called the *metric topology*.

(6) Bringing these concepts together we have a *normed topological vector space*. When such a space is complete it is called a Banach space.

(7) A Banach space on which an inner product has been defined and where the norm is derived from the inner product is called a *Hilbert space*. (See Chapter 3).

Comment

We are accustomed from an early age to visualize problems in Euclidean space which is simply a finite dimensional Hilbert space. Thus, in Hilbert space, geometric intuition often gives valuable insight. It is in the spaces with little structure that visualization is usually most difficult.

2.12 LINEAR FUNCTIONALS

A bounded linear mapping that maps a normed space X into the scalar field K is called a *bounded linear functional f*: $X \to K$. Let $X = C[a, b]$ be the space of all real valued functions continuous on an interval $[a, b]$. A familiar functional is then given by $f: X \to \mathbb{R}^1$ in the form of the integral $\int_a^b x(t)\, dt$. This functional is linear and is bounded if the interval $[a, b]$ is finite.

2.13 THE DUAL SPACE

Let X be a normed linear space and consider the set of all linear bounded functionals $f_1, f_2, \ldots, X \to K$ defined on X. By linearity we have that the sum of any two linear functionals is again a linear functional and that a scalar multiple of a linear functional is again a linear functional. On checking further, we find that the set of all linear bounded functionals on X is itself a linear space. It is called the *dual space of X*. A norm can be defined on the dual space, which is then denoted by X^*, i.e. X^* *is the normed dual space* of X. Let x^* be an element of X^* then the norm is defined as follows

$$\|x^*\| = \sup_{x \in \beta} \{|x^*(x)|\}$$

where β is the closed unit ball in X.

The dual space X^* is always complete since the scalar field is complete and the result follows by Corollary 2.2.1, given later in this chapter.

Hence, the dual space X^* is a Banach space, even if the original space X is not.

Since X^* is a linear space on which functionals can be defined it is natural to define the normed dual X^{**} of X^*, this is called the *second normed dual space* of X.

2.14 THE SPACE OF ALL BOUNDED LINEAR MAPPINGS

Let X, Y be normed spaces. Let $T: X \to Y$ be a linear mapping. First we show that for linear mappings boundedness and continuity are equivalent properties.

Theorem 2.1. *T is bounded if and only if T is continuous.*

Proof. Boundedness $\Rightarrow$ continuity. Assume boundedness and take two elements x_1, x_2 in the domain satisfying

$$\|x_1 - x_2\| < \delta, \qquad \delta > 0$$

Then

$$\|Tx_1 - Tx_2\| = \|T(x_1 - x_2)\| \leqslant M\|x_1 - x_2\| < M\delta$$

where M is the constant of boundedness for T. Put $\delta < \varepsilon/M$ then given δ defined above, there exists ε such that $\|Tx_1 - Tx_2\| < \varepsilon$ whenever $\|x_1 - x_2\| < \delta$, i.e. T is continuous.

Continuity $\Rightarrow$ boundedness. Assume that T is not bounded. There there exists a sequence $\{x_n\}$ such that

$$\|Tx_n\| > n\|x_n\|, \forall n$$

Put

$$z_n = \frac{x_n}{n\|x_n\|}$$

Then $\|z_n\| = 1/n$ so that $z_n \to 0$, $n \to \infty$. Now consider the mapping under

T of the sequence $\{z_n\}$

$$\{Tz_n\} = \left\{\frac{Tx_n}{n\|x_n\|}\right\}.$$

Then

$$\|Tz_n\| = \frac{\|Tx_n\|}{n\|x_n\|} \geqslant \frac{n\|x_n\|}{n\|x_n\|} = 1.$$

Since T is linear, $T(0) = 0$. Also $\lim_{n \to \infty} \{z_n\} = 0$, so that T is not continuous.

This contradiction shows that the original assumption was false. T is bounded and the proof is complete.

Let T_1, T_2, T be any linear mappings on the same linear space X into Y. Let $\alpha \in K$ then $(T_1 + T_2)$ and αT are again linear mappings on X. With addition and scalar multiplication defined in this obvious way the set of all bounded linear mappings from X to Y becomes a linear space, denoted $B(X, Y)$. (The reader should refer to the axioms for a linear space and check the truth of the assertion.)

In case the mappings are from X to itself, we follow the usual convention and denote the set of all bounded linear mappings $T: X \to X$ by $B(X)$. Thus $B(X, X) = B(X)$.

$B(X, Y)$ can be normed by defining

$$\|T\| = \sup\{\|Tx\| \, | x \in \beta(X)\}, \qquad T \in B(X, Y).$$

Theorem 2.2. *Let Y be a Banach space then the normed space $B(X, Y)$ is complete and hence is also a Banach space.*

Proof. Let $\{T_i\}$ be a convergent sequence of mappings. This means, more formally, that $\|T_n - T_m\| \to 0$ as $n, m \to \infty$, the norm being the operator norm. Choose some $x \in X$ then

$$\|T_n x - T_m x\| = \|(T_n - T_m)x\| \leqslant \|T_n - T_m\| \, \|x\| \to 0$$

as $n, m \to \infty$. Therefore, the sequence $\{T_i x\}$ is convergent for every x and since Y is complete, being a Banach space, the sequence has its limit in Y. Let the limit in Y be denoted y then we can write $y = Tx$—this equation defining T. T is easily shown to be linear and bounded, hence it is an element in $B(X, Y)$.

Finally, we show that T is the limit in $B(X, Y)$ of the sequence $\{T_i\}$. Choose $\varepsilon > 0$ then $\exists N$ such that $\|T_{n+p} x - T_n x\| < \varepsilon$, for all $n > N$, for all $p > 0$ and for all x in the closed unit ball of X. Let p go to infinity then we have that

$$\|Tx - T_n x\| < \varepsilon, \qquad n > N, x \text{ in the closed unit ball of } X.$$

Hence for $n > N$,

$$\|T - T_n\| = \sup_{x \in \beta(X)} \|(T - T_n)x\| = \sup_{x \in \beta(X)} \|Tx - T_n x\| < \varepsilon$$

so that $T = \lim_{n \to \infty} \{T_n\}$.

We have shown that the limit of a Cauchy sequence of mappings in $B(X, Y)$ exists and belongs to the space $B(X, Y)$ provided that the space Y is complete.

Corollary 2.2.1. *The space* $X^* = B(X, K)$ *of all bounded linear mappings from a linear space* X *to the scalar field* K *is a Banach space.*

Proof. K is complete and the result follows. (Recall that $B(X, K)$ is the set of all linear bounded functionals on X, usually denoted X^*).

2.15 EXERCISES

(1) Let E be a set with the discrete topology. Prove that every function defined on E is continuous.

(2) Show that the set of real numbers x in $\mathbb{R}^1$ satisfying $|x| > \pi$ is open in the usual topology for $\mathbb{R}^1$.

(3) Define a function on $X \times Y \to \mathbb{R}^1$

$$f(x, y) = 0 \qquad \text{if } x = y = 0$$

$$f(x, y) = \frac{x^2 - y^2}{x^2 + y^2} \qquad \text{otherwise}$$

Show that

$$\lim_{x \to 0} \left(\lim_{y \to 0} f(x, y) \right) \neq \lim_{y \to 0} \left(\lim_{x \to 0} f(x, y) \right)$$

Comment on this situation.

(4) Is it possible that a function f can be defined on an interval $[0, 1]$ such that

  (i) $f(t) \neq 0$ wherever $t = 1/i, \quad i = 1 \dots$.
  (ii) $\int_0^1 |f(t)| \, dt = 0$.

If such a function could be found, what would be a good general way in which to describe it?

(5) Let E be the set of real valued functions continuous on an interval

$[a, b]$. Show that

$$d(f_1, f_2) = \sup_{x \in [a,\, b]} \{|f_1(x) - f_2(x)|\}$$

satisfies the requirements for a metric.

(6) Prove as many as possible of the inclusion relations between sequence spaces (given in Section 2.9).

(7) Prove as many as possible of the inclusion relations between function spaces on a finite interval (given in Section 2.10).

(8) Prove that every finite dimensional normed linear space is a Banach space.

(9) Show that the l^p spaces, $p \in [1, \infty)$ are complete.

(10) Let $T: X \to Y$ be a bounded linear mapping between normed linear spaces. Prove that N_T is a closed subspace of X. Let X be a real normed linear space. Prove that a functional $f \in X^*$ is continuous if and only if N_f is closed.
  Hint: To prove that if N_f is closed then f is continuous. Suppose that f is not the zero functional (otherwise the result follows trivially). Then there exists $a \in X$ such that $a \notin N_f$. Since N_f is closed, there is a ball centred on a which does not meet N_f. Show by contradiction that f is bounded on this ball and hence that f is continuous.

(11) τ_1, τ_2 are topologies for a set X. Define τ_3 by $\tau_3 = \tau_1 \cap \tau_2$.
  Is τ_3 a topology for X?

(12) Prove that
  (i) Any closed subspace of a compact space is compact.
  (ii) The image of a compact set under a continuous mapping is compact.

CHAPTER 3

Inner Product Spaces and Some of their Properties

3.1 INNER PRODUCT

In applied mathematics, we are used to considering quantities in $\mathbb{R}^3$ like $|x| \cdot |y| \cos \theta$ where x, y are vectors and θ is the angle between x and y. A generalization of this concept is the inner product. The *inner product* of two elements x, y of a real or complex linear space Y is denoted $\langle x, y \rangle$ and has the following properties:

(i) $\langle x, x \rangle > 0$ whenever $x \neq 0$
(ii) $\langle x, y \rangle = \overline{\langle y, x \rangle}$
(iii) $\langle \alpha x, y \rangle = \alpha \langle x, y \rangle$, α a scalar
(iv) $\langle x + y, z \rangle = \langle x, z \rangle + \langle y, z \rangle$.

Any space on which an inner product has been defined is called an *inner product space*.

The inner product can immediately be used to define a norm by setting $\|y\| = \langle y, y \rangle^{1/2}$. The axioms for a norm can be seen to appear in the definition of the inner product except for the triangle inequality which is not difficult to prove.

3.2 ORTHOGONALITY

In an inner product space, Y, two elements $x, y \in Y$ are defined to be *orthogonal*, (denoted $x \perp y$) if $\langle x, y \rangle = 0$. Let Z be a subset of Y and y be an element of Y then y is said to be orthogonal to Z (denoted $y \perp Z$) if $\langle y, z \rangle = 0$, $\forall z \in Z$.

Let $\{y_i\}$ be a set of elements in an inner product space Y satisfying $\langle y_i, y_j \rangle = 0$, $i \neq j$, i.e. $\{y_i\}$ is a set of mutually orthogonal elements.

Define $e_i = y_i / \|y_i\|$, then each e_i has unit norm and the set $\{e_i\}$ is called an *orthonormal set*.

Recall that Y has finite dimension if a subset E of Y exists, called a basis for Y, such that every element of Y can be written uniquely as a finite linear combination of elements of E.

Inner product spaces will often have infinite dimension and the Schauder basis is a suitable extension of the (purely algebraic) concept of basis defined for finite dimensional spaces. Let Y be a linear space and let E be a sequence of elements $\{e_i\}$ of elements of Y.

E is called a *Schauder basis* for Y if every element $y \in Y$ can be uniquely represented by a sum $y = \sum_i \alpha_i e_i$, $\alpha_i \in K$.

Finally suppose that a sequence $\{e_i\}$ is a Schauder basis for Y while also $\{e_i\}$ is an orthonormal set, then $\{e_i\}$ is called an *orthonormal basis* for Y.

3.3 HILBERT SPACE

Let X be a linear space on which an inner product has been defined, with a norm derived from the inner product and which is complete in this norm. Then X is called a *Hilbert space*. More briefly X is a Banach space with an inner product from which its norm has been derived.

A general comment which should not be interpreted too literally is that geometric intuition usually leads to correct conclusions in Hilbert space (although it usually fails in the more general setting of a Banach space). This makes generalization of familiar finite dimensional theorems easy to visualize, with the projection theorem which follows later being a good example.

Examples of Hilbert spaces

(1) $\mathbb{R}^n$ is a Hilbert space with inner product

$$\langle x, y \rangle = \sum_{i=1}^{n} x_i y_i$$

(2) l^2 is a Hilbert space with inner product

$$\langle x, y \rangle = \sum_{i=1}^{\infty} x_i y_i$$

(3) L^2 is a Hilbert space with inner product

$$\langle x, y \rangle = \int_I x(t) y(t) \, \mathrm{d}t$$

where the functions x, y are defined and integrable on a domain I.

3.4 THE PARALLELOGRAM LAW

Let X be a Hilbert space and x, y be two elements of X. Then

$$\|x + y\|^2 = \langle x + y, x + y \rangle$$
$$= \langle x, x \rangle + \langle y, x \rangle + \langle x, y \rangle + \langle y, y \rangle$$
$$= \|x\|^2 + \overline{\langle x, y \rangle} + \langle x, y \rangle + \|y\|^2$$
$$= \|x\|^2 + 2\,\mathscr{R}(\langle x, y \rangle) + \|y\|^2.$$

Similarly

$$\|x - y\|^2 = \|x\|^2 - 2\mathscr{R}(\langle x, y \rangle) = \|y\|^2$$

so that

$$\|x + y\|^2 + \|x - y\|^2 = 2\|x\|^2 + 2\|y\|^2$$

where $x + y$ is the long diagonal and $x - y$ the short diagonal of a parallelogram.

3.5 THEOREMS

Theorem 3.1. (Existence of a unique minimizing element.) *Let X be a Hilbert space and M be a closed subspace of X. Let $x \in X$, $x \notin M$ then there exists an element $y \in M$ and a real number d satisfying*

$$\|x - y\| = \inf_{z \in M} \|x - z\| = d$$

Proof. Let $\{z_i\}$ be a sequence in M such that

$$\|x - z_i\| \to d$$

Let z_m, z_n be any two elements of the sequence $\{z_i\}$ then by the parallelogram law

$$\|(x - z_n) + (x - z_m)\|^2 + \|(x - z_n) - (x - z_m)\|^2 = 2\|x - z_n\|^2 + 2\|x - z_m\|^2$$

$$4\left\|x - \frac{z_n + z_m}{2}\right\|^2 + \|z_m - z_n\|^2 = 2\|x - z_n\|^2 + 2\|x - z_m\|^2 \quad (3.5.1)$$

$\frac{1}{2}(z_n + z_m) \in M$ since M is a subspace.

Thus,

$$\|x - \tfrac{1}{2}(z_n - z_m)\| \geqslant d$$

and equation (3.5.1) can be rewritten

$$4d^2 + \|z_m - z_n\|^2 \leqslant 2\|x - z_n\|^2 + 2\|x - z_m\|^2$$

$$\tfrac{1}{4}\|z_m - z_n\|^2 \leqslant \tfrac{1}{2}\|x - z_n\|^2 + \tfrac{1}{2}\|x - z_m\|^2 - d^2$$

Let $m, n \to \infty$ then the right-hand side goes to zero. Hence $\{z_i\}$ is a Cauchy sequence in a closed subspace of a complete space, and hence

$$\lim_{i \to \infty} \{z_i\} = y \in M$$

(Since the limit of the sequence must be unique.)

Theorem 3.2. (Unique decomposition property or projection theorem.) *Let M be a closed subspace in a Hilbert space X, then every $x \in X$ can be uniquely decomposed as $x = v + w$, $v \in M$, $w \in M^{\perp}$.*

Proof. If $x \in M$ then trivially $v = x$, $w = 0$. Thus, assume $x \notin M$. Choose $y \in M$ such that

$$\|x - y\| = \inf_{z \in M} \|x - z\| = d$$

(see previous theorem). Now let $v = y$ and $w = x - y$. Let z be any non zero element in M. Let α be any scalar then

$$d^2 \leqslant \|w + \alpha z\|^2 = \|w\|^2 + 2\alpha\langle w, z \rangle + \alpha^2 \|z\|^2$$

$$= \|z\|^2 \left(\alpha + \frac{\langle w, z \rangle}{\|z\|^2}\right)^2 + d^2 - \frac{\langle w, z \rangle^2}{\|z\|^2}$$

Now put $\alpha = -\langle w, z \rangle / \|z\|^2$ then $\langle w, z \rangle^2 \leqslant 0$. Thus w is orthogonal to z and we have proved that $w \in M^{\perp}$ while since $v = y$, $v \in M$.

To show uniqueness assume that

$$x = v + w = v' + w'; \qquad v, v' \in M, \qquad w, w' \in M^{\perp}$$

Then

$$v - v' = w' - w = q \text{ (say)}$$

The element $q = v - v'$, hence q is in M (subspace property). Also $q = w' - w$, hence q is in $M^\perp$. Only $q = 0$ can satisfy these conditions, hence $q = 0$ and uniqueness is proved.

Corollary 3.2.1. *There exists an element* $w \neq 0$ *in* X *which is orthogonal to* M.

Proof. Let $x \in X$, $x \notin M$ then from Theorem 3.2. we can write $x = v + w$, $v \in M$, $w \in M^\perp$.

Comment

Referring back to Theorem 3.1, it is clear that $(x - y) \in M^\perp$ while $y \in M$. Hence $(x - y)$ is orthogonal to y. This can be considered a corollary to Theorem 3.1.

Theorem 3.3. (Riesz representation theorem). *Let* X *be a Hilbert space and take any bounded linear functional* $f : X \to K$. *Then there exists some element* $g \in X$ *such that* $f(x) = \langle x, g \rangle$ *(i.e. the functional can be represented by an inner product). Further* $\| f \| = \| g \|$. *(It is clear that if* g *is fixed in the inner product* $\langle x, g \rangle$ *then the inner product is a linear functional* $X \to K$. *The theorem asserts the converse, that every functional on* X *can be represented by an inner product.)*

Proof. If f maps every element of X into zero then we take $g = 0$. In the other cases g must be non-zero. Let N_f be the null space of f, i.e.

$$N_f = \{x \,|\, f(x) = 0\} \quad \text{then} \quad x \in N_f \Rightarrow \langle x, g \rangle = 0$$

i.e.

$$g \perp N_f$$

Now N_f is a subspace of X (check this) and N_f is closed since, let $\{x_i\} \to x$ be a sequence in N_f then

$$f(x) = f(x) - f(x_i) \quad \text{since} \quad f(x_i) = 0$$
$$= f(x - x_i)$$

Thus

$$|f(x)| = |f(x - x_i)| \leqslant k \| x - x_i \| \quad \text{for some} \quad k \in K$$

The right-hand side goes to zero as $i \to \infty$, hence $f(x) = 0$, $x \in N_f$ and N_f is closed.

Now let z be any element in X satisfying $z \notin N_f$ then by Theorem 3.2 we can write

$$z = v + g, \qquad v \in N_f, \qquad g \in N_f^\perp$$

Thus, there does exist a g satisfying $g \perp N_f$ as required. g is unique since let g' satisfy

$$\langle x, g \rangle = \langle x, g' \rangle$$

then

$$\langle x, g - g' \rangle = 0, \quad \forall x, \quad \text{i.e. } g = g'$$

Finally, by definition

$$\|f\| = \sup_{\substack{x \in X \\ x \neq 0}} |f(x)| / \|x\|$$

Put $x = y$, then $|f(x)| / \|x\|$ takes on its supremum and becomes

$$|\langle y, y \rangle| / \|y\| = \|y\|$$

Thus

$$\|f\| = \sup_{x \in X} |f(x)| / \|x\| = \|y\|$$

Let E be a Hilbert space with an orthonormal basis $\{\alpha_i\}$. Let $T: E \to E$ be a bounded linear operator. Then T can be represented by a matrix A with coefficients $a_{ij} = \langle \alpha_i, T\alpha_j \rangle$.

Clearly, if the space E is finite dimensional, the study of operators can be entirely in terms of matrices. In the general case the operator T can be considered to be represented by a sequence of Fourier coefficients. Unbounded operators are encountered in the study of distributed parameter systems. However, if the operator T is unbounded, then there is no result equivalent to the above. If the operator is unbounded but closed, useful results can still be found. Such results are given and used in the chapter on distributed parameter systems.

3.6. EXERCISES

(1) Show that $\langle x, y \rangle$ in $\mathbb{R}^n$ satisfies the given axioms for an inner product and show also that $\langle x, y \rangle^{1/2}$ satisfies the given axioms for a norm.

(2) Let X be a Hilbert space and let $x, y \in X$. Prove the Schwarz inequality

$$\langle x, y \rangle \leqslant \|x\| \|y\|$$

Hint: $\langle x - \alpha y, x - \alpha y \rangle \geqslant 0$ for any $\alpha \in K$: put $\alpha = \langle x, y \rangle / \langle y, y \rangle$.

(3) Let X be a Hilbert space and let $x, y \in X$ such that $\langle x, y \rangle = 0$. Prove that

$$\|x + y\|^2 = \|x\|^2 + \|y\|^2$$

(known as the Pythagorean law).

(4) Show that the norm in Hilbert space satisfies the parallelogram law.

(5) Let E be the set of functions in $X = L^2[a, b]$ defined by

$$E = \{f \mid f \text{ continuous on } [a, b]\}$$

Is E closed in x?

(6) Let X be an infinite dimensional vector space. Let $\{e_i\}$, $i = 1, \ldots$, be an infinite set of linearly independent vectors in X. Does $\{e_i\}$ necessarily form a basis for x?

(7) Show that the function $\sin(nx)$ is orthogonal to each of the functions $\cos(nx)$, $\sin(mx)$, $\cos(mx)$, $m \neq n$, on the interval. $(-\pi, \pi)$ (m, n being integers). Go on to show that the set of functions

$$\left\{\frac{1}{\sqrt{\pi}} \cos(nx) \frac{1}{\sqrt{\pi}} \sin(nx)\right\}, \qquad n = 1, \ldots,$$

is an orthonormal set on $[-\pi, \pi]$.

(8) Let X be a Hilbert space and let $f \in X$, let $\{e_i\}$ be an orthonormal set in X. Define

$$g_n = \sum_{i=1}^{n} \alpha_i e_i$$

Define

$$J_n = \|f - g_n\|^2 = \langle f, f \rangle - 2\langle f, g_n \rangle + \langle g_n, g_n \rangle$$

show that J is minimized when

$$\alpha_i = \langle f, e_i \rangle, \qquad i = 1, \ldots, n$$

The α_i are the Fourier coefficients of f.

CHAPTER 4

Some Major Theorems of Functional Analysis

4.1 INTRODUCTION

In this chapter we give (without proofs) some of the major theorems of functional analysis. These are:

The Hahn–Banach theorem and its geometric equivalent
The closed graph theorem.
The Banach inverse theorem.
The open mapping theorem.
The uniform boundedness theorem.

The Hahn–Banach theorem has extensive application in later chapters and should be fully understood in its several interpretations. The other theorems will not be specifically used in control applications.

Two other topics are treated in this chapter:

Hölder's inequality—used extensively in manipulation of control problems in a concrete setting.
Norms on product spaces—required in the extension of results to spaces of higher dimension.

4.2 THE HAHN–BANACH THEOREM AND ITS GEOMETRIC EQUIVALENT

Theorem 4.1. (The Hahn–Banach theorem). The Hahn–Banach theorem is the fundamental tool to be used in establishing the existence of optimal controls. The theorem asserts that:

Given a normed space X and a subspace M ⊂ X with a bounded linear functional f defined on M, there exists a bounded linear functional g, defined on the space X and satisfying

(i) $g(x) = f(x), \forall x \in M$.

(ii) $\|g\| = \|f\|$ (*norm g being taken over X and norm f over the subspace M*).

The theorem means that a functional defined on a subspace can be extended to the whole space with preservation of norm.

Let X be any normed space and choose an arbitrary element $x \in X, x \neq 0$. Let the set $M = \{\lambda x \,|\, \forall \lambda \in K\}$. Now define a functional f on M by $f(\lambda x) = \lambda \|x\|$. It is clear that this is well defined. By the Hahn–Banach theorem, f can always be extended to the whole space X. Thus, at least one nonzero continuous functional exists on the whole of X, hence the usefulness of the theorem in existence proofs (but note that the functional g is not necessarily unique).

Geometric aspects

Let E be a linear space and let N be a linear subspace of E. Let

$$x \in E, \qquad x \neq 0$$

then $V = x + N$ is the translation of a linear subspace and V is called a *linear manifold* or linear variety.

A *hyperplane* is a linear manifold V that is maximal, i.e. such that:

(i) $V \neq E$.

(ii) $V \subset W \subset E$, W a linear manifold $\Rightarrow W = E$ or $W = V$.

Let H be a hyperplane; then there exists a nonzero functional f and a scalar λ such that

$$H = \{x \,|\, f(x) = \lambda\}$$

with f, λ being unique up to proportionality. A hyperplane H in a normed space is closed if and only if its corresponding functional is continuous.

Let E be a real linear space and let K be a convex subset of E. A hyperplane H in E is said to *support* K if:

(i) K belongs to one of the closed half spaces generated by H.

(ii) $K \cap H \neq \varnothing$.

Then there exists a nonzero functional g and a real number α such that

$$g(x) \leqslant \alpha \text{ on } K \text{ with } g(x) = \alpha \text{ for at least one } x \in K.$$

Let E be a real normed linear space and let K be a convex subset of E then:

(i) Every supporting hyperplane of K is closed.

(ii) Every $x \in \partial K$ is contained in a supporting hyperplane H.

A point x is an *extreme point* of a set K if $x \in K$ cannot be expressed as a proper combination of two distinct elements of K. If K is convex, then $x \in K$ is an extreme point of K if x cannot be written as the midpoint of any line segment in K.

A point $y \in K$ is an *exposed point* of K if there exists a closed supporting hyperplane H for K satisfying

$$H \cap K = \{y\}$$

Mazur's theorem

Mazur's theorem is the geometric version of the Hahn–Banach theorem. As has been stated above, the linear functionals can be identified with hyperplanes. Accordingly, Mazur's theorem asserts the existence of a hyperplane. The theorem is often valuable in giving geometric insight.

Theorem 4.2. (Mazur's theorem). *Let A, B be convex subsets of a normed space X. Let*

$$B^0 \neq \varnothing, \qquad B^0 \cap A = \varnothing$$

then A and B can be separated by a hyperplane or equivalently, there exists a linear functional f, not identically zero, such that

$$\sup_{x \in A} f(x) \leqslant \inf_{x \in B} f(x)$$

Alternative form

Let K be a convex subset of X and let $x \in \partial K$. Let $K^0 \neq \varnothing$, then there exists a hyperplane supporting K at x, x being known as a supporting point *or, equivalently, there exists a nonzero functional on X with $f(x) = \sup f(K)$. We call f a support functional.*

4.3 OTHER THEOREMS RELATED TO MAPPINGS

Many useful theorems require that the linear mappings that play a major role in the representation of systems shall be bounded. In some applications, unbounded mappings are encountered but provided that the mappings are closed, useful results, analogous to those for bounded mappings can still be drawn.

Let A be a linear mapping on a normed space X into Y. A is defined to be a *closed mapping* if given a sequence $\{x_i\}$ in X we have that

$$\{x_i\} \to x \quad \text{and} \quad \{Ax_i\} \to y, \quad \forall x \in X \Rightarrow y = Ax$$

Clearly, every continuous linear mapping is closed, hence bounded. Distinguish carefully between a closed mapping and a continuous mapping by recalling that A is continuous if

$$\{x_i\} \to x \Rightarrow \{Ax_i\} \to y = Ax$$

Thus, A is a closed mapping if y exists and is equal to Ax. For a continuous mapping, y always exists and is equal to Ax.

If A is a closed mapping and A^{-1} exists then A^{-1} is also a closed mapping. This is not a trivial result since the inverse of a bounded mapping is not necessarily bounded.

The *graph* of a mapping $A: X \to Y$ denoted G_A is defined as the set of ordered pairs

$$G_A = \{(x, y) | (x, y) \in X \times Y, y = Ax\}.$$

Theorem 4.3. (The closed graph theorem). *Let $A: X \to Y$ be a mapping between the Banach spaces X and Y. Let A be linear. Then A is bounded $\Leftrightarrow G_A$ is closed.*

When considering closed mappings, the domain must always be carefully defined, for by the closed graph theorem:

Let A be a closed linear mapping defined on the whole of a Banach space X then A cannot be unbounded, i.e. A is necessarily continuous.

Thus, every unbounded closed mapping will be defined on a restricted domain.

Theorem 4.4. [The Banach inverse theorem (bounded inverse theorem)]. *Let X, Y be Banach spaces and let A be a bounded linear mapping from X onto Y. If A^{-1} exists, then A^{-1} is a bounded linear mapping from Y to X, i.e.*

$$A \in B(X, Y) \Rightarrow A^{-1} \in B(Y, X).$$

The theorem is a consequence of the closed graph theorem for if A^{-1} exists it is a closed mapping from Y to X with the domain of A^{-1} being the whole of Y. Hence A^{-1} is bounded.

Finally, we note that A need not be bounded but merely closed for the theorem to apply.

Theorem 4.5. (The open mapping theorem). *Let U, Z be Banach spaces and let S be a surjective bounded linear mapping from U onto Z*

$$S: U \to Z$$

Then S is an open mapping such that if G is open in U, $S(G)$ is open in Z.

Alternative statements of the theorem are:

(i) *If S is a bounded linear mapping with closed range then S is an open mapping.*

(ii) *Let β_u, β_z be the unit balls in U and Z then $\alpha\beta_z \subset S\beta_u$ for some real $\alpha > 0$.*

(iii) *If $z_1 \in Z$ with $\|z_1\| \leqslant \alpha > 0$ then there exists at least one u_1 with $\|u_1\| \leqslant 1$ such that*

$$Su_1 = z_1$$

Theorem 4.6. [The uniform boundedness theorem (Banach–Steinhaus theorem)].
Let X be a Banach space, Y a normed linear space and W any subset of $B(X, Y)$ such that

$$\inf_{A \in W} \|Ax\| < \infty, \qquad \forall x \in X$$

Then there exists a constant M such that

$$\|A\| \leqslant M, \qquad \forall A \in W.$$

4.4 HÖLDER'S INEQUALITY

For some control problems, it is possible to write down an inequality that must be satisfied if any sort of control is to be achieved. If the control is to be optimal then the inequality must be satisfied with equality. The most useful inequality for our work is Hölder's.

Let $1 < p < \infty, q = p/(p - 1)$. Let $x \in L^p, f \in L^q$ then the pointwise product $f.x \in L^1$ and

$$\left| \int_I x(t)f(t) \, dt \right| \leqslant \left[\int_I |x(t)|^p dt \right]^{1/p} \left[\int_I |f(t)|^q \, dt \right]^{1/q.}$$

and if $x \in l^p, f \in l^q$ then

$$\left| \sum x_i f_i \right| \leqslant [|x_i|^p]^{1/p} [|f_i|^q]^{1/q}$$

Conditions for equality in the Hölder inequality

From the above

$$\left| \int_I x(t)f(t) \, dt \right| \leqslant \int_I |x(t)f(t)| \, dt \leqslant \left[\int_I |x(t)|^p \, dt \right]^{1/p} \left[\int_I |f(t)|^q \, dt \right]^{1/q}$$

The first half of the inequality will be satisfied with equality if and only if

or
$$\left.\begin{array}{l} \text{sign}(x(t)) = \text{sign}(f(t)) \\ \text{sign}(x(t)) = -\,\text{sign}(f(t)) \end{array}\right\} \quad \forall\, t \in I$$

The second inequality becomes an equality if and only if

$$|x(t)| = k|f(t)|^{q-1}, \qquad \forall\, t \in I$$

combining the two conditions leading to the condition for equality in

$$x(t) = k|f(t)|^{q-1}\,\text{sign}(f(t))$$

When X is an l^p space the inequality chain below applies:

$$\left|\sum x_i f_i\right| \leqslant \sum |x_i f_i| \leqslant \left[\sum |x_i|^p\right]^{1/p}\left[|f_i|^q\right]^{1/q}$$

The first inequality becomes an equality if

$$\text{sign}(x_i) = \text{sign}(f_i)$$

The second inequality becomes an equality if

$$|x_i| = |k|\,|f_i|^{q-1}$$

while the original inequality is satisfied with equality if

$$x_i = k|f_i|^{q-1}\,\text{sign}(f_i), \qquad \forall i,\ k \text{ an arbitrary constant}$$

If we have integrals of the form

$$\int_I \sum x_i(t) f_i(t)\,dt$$

Hölder's inequality can be extended to yield

$$\int_I \sum |x_i(t) f_i(t)|\,dt \leqslant \int_I \left(\sum |x_i(t)|^p\right)^{1/p}\left(\sum |f_i(t)|^q\right)^{1/q}\,dt$$

with equality holding if and only if

$$x_i(t) = k|f_i(t)|^{q-1}\,\text{sign}(f_i(t))$$

In these equality conditions, if $x \in X$ is required to satisfy $\|x\| = 1$ then k takes on a particular value, when X is an L^p space, the condition for equality takes the form

$$\left|\int_I k|f(t)|^{q-1}\,\text{sign}(f(t)).\,f(t)\,dt\right| = \left(\int_I |f(t)|^q\,dt\right)^{1/q}$$

$$\left| k \int_I |f(t)|^{q-1} |f(t)| \, dt \right| = \left(\int_I |f(t)|^q dt \right)^{1/q}$$

$$k \int_I |f(t)|^q \, dt = \left(\int_I |f(t)|^q dt \right)^{1/q}$$

$$k = \frac{(\int_I |f(t)|^q \, dt)^{1/q}}{\int_I |f(t)|^q \, dt} = \frac{\|f\|}{\|f\|^q} = \|f\|^{1-q}$$

yielding the condition to be satisfied by x in order that $f(x)$ shall assume its supremum on the unit ball in X as

$$x(t) = \|f\|^{1-q} |f(t)|^{q-1} \operatorname{sign}(f(t))$$

Similarly for X an l^p space the condition is

$$x_i = \|f\|^{1-q} |f_i|^{q-1} \operatorname{sign}(f_i)$$

Given a particular functional f on a Banach space X, $f \in X^*$ these relations allow a particular $x \in \beta \subset X$ to be chosen, if it exists, such that $f(x)$ takes on its supremum.

From the symmetry of the functional $f(x)$ which can be written $\langle x, f \rangle$ it is obviously possible to consider $x \in X$ as fixed and to seek the element $f \in \beta^* \subset X^*$ that causes $f(x)$ to take on its supremum.

4.5 NORMS ON PRODUCT SPACES

In the optimal control of systems with several inputs and several states it is necessary to arrive at a scalar valued function whose value must be minimized or maximized. Such a scalar valued function is provided by the norm on a suitable product space.

Let $\{X_i\}, i = 1, \ldots, n$ be a family of normed linear spaces and let $X = \prod^n X_i$ be a product space. If each member of the family $\{X_i\}$ is separately normed then given any $x = (x_1, \ldots, x_n) \in X$, we can define an element $r \in \mathbb{R}^n$ by the relation

$$r(x) = (\|x_1\|_1, \ldots, \|x_i\|_i, \ldots, \|x_n\|_n)$$

Then *any* norm for $\mathbb{R}^n$ can be used on $r(x)$ and the result is a norm for the product space X. Typically,

$$\|x\| \triangleq \left(\sum_{i=1}^{n} \|x_i\|^p \right)^{1/p}, \qquad 1 \leqslant p \leqslant \infty.$$

4.6 EXERCISES

(1) Prove Mazur's theorem for the restricted case where the convex body is the closed unit ball β in a real normed linear space X.

Hint: Let $p(x) = \|x\|$ be defined on X. Show that there exists a functional f satisfying $f(x) \leqslant p(x)$. By the Hahn–Banach theorem f can be extended to a functional F and the hyperplane sought is the translate of the null space of F.

(2) Let X be a normed space. Let $f \in X^*$. Show that the set

$$\{x \mid f(x) = c, x \in X\}$$

is a hyperplane in X.

(3) Let X be a normed linear space and H a hyperplane in X. Prove that H is either closed in X or is dense in X. Prove that if H is closed, then the functional corresponding to the hyperplane is continuous. (See Problem 10, Section 2.15.)

(4) Let $A: X \to Y$ be a linear continuous mapping between normed spaces. Show that if A^{-1} exists then it is linear.

Is it necessarily true that A^{-1} will be continuous?

(5) Let X be the set of functions differentiable on $[a, b]$. Let $T: X \to C[a, b]$ be the differential operator $T: x \to \dot{x}$. Show that T is closed but not continuous.

(6) Prove Hölder's inequality (Section 4.4) and illustrate its use in a function space context.

CHAPTER 5

Linear Mappings and Reflexive Spaces

5.1 INTRODUCTION

In this chapter are collected most of the results from functional analysis that will be used explicitly in applications to the study of control systems. The material on reflexive spaces and their properties is often called geometric functional analysis and has a fairly obvious geometric interpretation that is a great advantage in visualization.

We shall see later that dynamic systems can be represented by mappings between appropriate spaces. For systems governed by sets of ordinary differential equations, the range of the mapping is always finite dimensional and the following theorems are useful.

5.2 MAPPINGS OF FINITE RANK

Let S be a linear mapping from U to X; U, X arbitrary normed spaces $S: U \to X$. If range S is finite dimensional then S is said to be a *mapping of finite rank*. In particular, if range S has dimension n, then the *rank of the mapping* is said to be equal to n.

Theorem 5.1. *Each linear mapping of finite rank can be expressed in terms of a summation involving n linearly independent functionals, where n is the rank of the mapping i.e. let $S: U \to X$ where S is a linear mapping of rank n then*

$$S(u) = \sum_{i=1}^{n} f_i(u)e_i$$

Proof. Dim Range $S \subset X = n$. Let $\{e_i\}$, $i = 1, \ldots, n$, be a basis for Range S. Choose $u \in U$ then $S(u)$ can be written $S(u) = \sum_{i=1}^{n} \alpha_i e_i$. The coefficients α are linear functionals of u and the theorem is proved.

5.3 MAPPINGS OF FINITE RANK ON A HILBERT SPACE

Theorem 5.2. *Let S be a linear mapping between a Hilbert space U and a finite dimensional space X, $S: U \to X$. Then there exists a finite dimensional space $M \subset U$ such that S restricted to M is an injective mapping.*

Proof. Let $\{e_i\}$, $i = 1, \ldots, n$, be a basis for range $S \subset X$. Given any $u \in U$, Su can be expressed, $Su = \sum \alpha_i e_i$, $i = 1, \ldots, n$ where each of the α_i can be expressed

$$\alpha_i = \langle f_i, u \rangle, \qquad f_i \in U^* = U$$

Then

$$Su = \sum_{i=1}^{n} \langle f_i, x \rangle e_i$$

The f_i are linearly independent and generate an n dimensional subspace, say $M \subset U$. Hence $U = M \oplus M^\perp$ from the properties of Hilbert space. Let $u \in M^\perp$ then $\langle f_i, u \rangle = \alpha_i = 0$. Thus $M^\perp \subset N_s$.
   Let $u \in N_s$ then

$$Su = 0 = \sum \langle f_i, u \rangle e_i$$

Since the e_i are independent, each $\langle f_i, u \rangle = 0$, $i = 1, \ldots, n$, so that $M^\perp = N_s$. S maps M bijectively to range S and hence S restricted to M is an injective mapping into X.

5.4 REFLEXIVE SPACES

Those spaces where the space X and its second dual are isomorphic are particularly important.

A normed linear space X is defined to be *reflexive* if $X = X^{**}$. Since X^{**} is always complete, every reflexive space is a Banach space. Let X be a reflexive space then two important properties are:

   (i) The unit ball $B \subset X$ is weak compact—this fact can be used in weak analogues of theorems where compactness is required (weak compactness is defined later in this chapter).
   (ii) Every decreasing sequence of closed bounded convex sets has non-empty intersection.

Lemma 5.3. *Let U, Y be Banach spaces and let $R: U \to Y$ be a bounded linear mapping. If U is a reflexive space then the graph of $R = G_R = \{(u, Ru) | u \in U\}$ is also a reflexive space.*

Proof. The lemma is proved if the canonical mapping $Q: G_R \to G_R^{**}$ can be shown to be surjective.

Define a mapping $H: U \to G_R$ by $Hu = (u, Ru)$. H is linear and bijective. $H^{**}: U^{**} \to G_R^{**}$ is also linear and bijective. The mapping $Q': U \to U^{**}$ is bijective by definition since U is reflexive.

Hence $Q^{-1} = H(Q')(H^{**})^{-1}$ is a left inverse of Q.

Thus $Q = H^{**}Q'H^{-1}$, and since all three mappings are bijective, Q is also bijective and G_R is reflexive. (Q^{-1} is a *left inverse* of Q if $Q^{-1}Q = I$.)

5.5 ROTUND SPACES

Let β be the unit ball in a Banach space U and let $\partial\beta$ be the boundary of β. The space U is defined to be *rotund* if one of the following equivalent conditions is satisfied.

(i) $\|x_1 + x_2\| = \|x_1\| + \|x_2\| \Rightarrow x_2 = \lambda_1 x_1$ for some scalar $\lambda_1 \neq 0$.

(ii) Each convex subset K of U has at most one element satisfying
$$\|u\| \leqslant \|z\|, \quad u \in K, \quad \forall z \in K.$$

(iii) For any bounded linear functional f on U there is at most one $x \in \beta$ such that
$$\langle x, f \rangle = f(x) = \|f\|$$

(iv) Every $x \in \partial\beta$ is an extreme point of β.

(v) Each hyperplane of support meets β in exactly one point.

Examples of rotund spaces

(1) Every Hilbert space is rotund.

(2) Every l^p, L^p space is rotund for $1 < p < \infty$.

(3) Let $U_1, \ldots, U_n$ be rotund Banach spaces, then the space $U_1 \times U_2 \times \ldots \times U_n$ is also rotund provided that a norm of the form
$$\|u\| = \left(\sum_i^n \|x_i\|^p \right)^{1/p}, \quad 1 < p < \infty$$
is used.

(4) Every uniformly convex space is rotund. (Every locally compact rotund space is uniformly convex).

Comment

There is no known example of a reflexive space that is not rotund.

5.6 SMOOTH SPACES

A normed linear space will be called *smooth* if at each point $x \in \partial\beta$ there exists one and only one hyperplane.

Equivalently, a space X is smooth if for every $x \in X$, $x \neq 0$ there exists at most one functional f such that $f(x) = \|x\|$.

Rotundity and smoothness are closely related; in fact, rotundity of $X^* \Rightarrow$ smoothness of X, while if the space X is reflexive then rotundity of $X^* \Leftrightarrow$ smoothness of X, and smoothness of $X^* \Leftrightarrow$ rotundity of X.

5.7 UNIFORM CONVEXITY

A normed space X is *uniformly convex* if given $\varepsilon > 0$, $\exists \delta$ such that

$$\|x_1 - x_2\| > \varepsilon \Rightarrow \tfrac{1}{2}\|x_1 + x_2\| < 1 - \delta, \qquad x_1, x_2 \in \beta$$

Every uniformly convex space is rotund but the converse is generally not true. However, every finite dimensional rotund space is uniformly convex. Rotundity requires that

$$\tfrac{1}{2}\|x_1 + x_2\| < 1, \qquad x_1, x_2 \in \beta, \qquad x_1 \neq x_2$$

while uniform convexity requires

$$\tfrac{1}{2}\|x_1 + x_2\| < 1 - \delta(\varepsilon)$$

where $\|x_1 - x_2\| = \varepsilon$, $\delta > 0$. Uniform convexity implies rotundity but not conversely.

Theorem 5.4. (Milman and Pettis theorem). *Every uniformly convex Banach space is reflexive but the converse is not true* (*see* D4 *for counterexamples*).

A space is uniformly convex if it is both rotund and locally compact. The L^p spaces are uniformly convex for $1 < p < \infty$.

5.8 CONVERGENCE IN NORM (STRONG CONVERGENCE)

A sequence of elements $\{f_i\}$, each $f_i \in L^p[a, b]$ *converges in norm* to a function $f \in L^p[a, b]$ if

$$\lim_{i \to \infty} \|f_i - f\| = \left(\int_a^b |f_i - f|^p \, dx \right)_{i \to \infty}^{1/p} = 0$$

When we speak simply of a convergent sequence of functions in a space X, we shall mean a sequence of functions convergent in the norm of X. (Such convergence is also called convergence in the natural topology of X or convergence in the mean (power p) or strong convergence.)

5.9 WEAK CONVERGENCE

A sequence $\{x_i\}$ in X is *weakly convergent* to an element $x \in X$ if for every $y \in X^*$,

$$\lim_{i \to \infty} \langle x_i, y \rangle = \langle x, y \rangle$$

When $X = l^2$, the classic example of a weakly convergent sequence is given by $\{x_i\}$ where $x_i = (0, 0, 0, 0, 1, 0, 0, \ldots)$ with unity in the ith place. We have $\langle x_i, y \rangle \to 0$ so that $\{x_i\}$ is weakly convergent but since $\|x_i - x_j\| = \sqrt{2}$, $\forall i \neq j$, $\{x_i\}$ is not convergent in the usual sense.

5.10 WEAK COMPACTNESS

A set E in X is weakly compact if any sequence in E has a subsequence that is weakly convergent to an element in E, i.e. let $\{x_i\}$ be a sequence in E, then there exists a subsequence $\{x_{i_k}\}$ and an $x \in X$ such that for all $f \in X^*$

$$\operatorname*{Lim}_{k \to \infty} f(x_{i_k}) = f(x)$$

5.11 WEAK* CONVERGENCE AND WEAK* COMPACTNESS

A sequence $\{f_i\}$ of elements in X^* is said to be *weak* convergent* to an element $f \in X^*$ if for every $y \in X$

$$\lim_{i \to \infty} \langle f_i, y \rangle = \langle f, y \rangle.$$

(It is a consequence of the Banach–Steinhaus theorem that a weak* convergent sequence is a bounded sequence.)

A set E in X^* is weak* compact if any sequence in E has a subsequence that is weak* convergent to an element in E, i.e. let $\{f_i\}$ be a sequence in E then there exists a subsequence $\{f_{i_k}\}$ and an $f \in X^*$ such that for all $x \in X$

$$\operatorname*{Lim}_{k \to \infty} f_k(x) = f(x)$$

5.12 WEAK TOPOLOGIES

Let X be any set and let Y be a topological space with topology, τ_y. Let $\{f_\alpha | \alpha \in Z\}$ be a collection of functions $f_\alpha : X \to Y$. The *weak topology* generated by $\{f_\alpha\}$ is the weakest topology such that each f_γ is continuous. The requirement is that $f_\alpha^{-1}(A)$ be open in X for each α where $A \in \tau_y$. Let $S = \{f_\alpha^{-1}(A) | \alpha \in Z, A \in \tau_y\}$ then S is a sub-base that will generate the weak topology, i.e. the collection of all finite intersections of elements of S forms a base for the weak topology.

Such a base consists of neighbourhoods $\{x \in X \mid \|f(x - x_0)\| < \varepsilon\}$. The weak topology of $\mathbb{R}^n$ is the same as the Euclidean metric topology. However, in an infinite dimensional Banach space the weak topology is not metrizable.

Theorem 5.5. *If K is a convex set that is closed in the weak topology, then K is closed in the norm topology—the converse is also true.*

The normed space X^* has two weak topologies. These are $\tau(X^*, X^{**})$ the *weak topology* for X^* and $\tau(X^*, X)$ which is called the *weak* topology* for X^*. $\tau(X^*, X)$ is produced by considering elements $x \in X$ as linear functionals on X^*. We note the facts.

(i) The weak* topology is weaker than the weak topology.
(ii) If X is reflexive the weak topology for X^* is the same as the weak* topology for X^*.
(iii) X^* with either the weak or the weak* topology is a topological vector space.

Many theorems in analysis such as the Weierstrass theorem below require compactness.

Theorem 5.6. (Weierstrass theorem). *A continuous functional on a compact subset K of a normed linear space X achieves its maximum on K.*

Let K be a compact subset of a normed space X. Given an arbitrary sequence $\{x_n\}$ in K there is a sub-sequence that is convergent to an element $x \in K$.

However, compactness is a very strong requirement and in infinite dimensional spaces the closed unit ball fails to be compact as is shown below.

5.13 FAILURE OF COMPACTNESS IN INFINITE DIMENSIONAL SPACES

Let β be the unit ball in l^2 and let $\{e_i\}$ be a sequence in β defined by

$$e_1 = (1, 0, 0, \ldots), \qquad e_2 = (0, 1, 0, \ldots)$$

This sequence contains no convergent sub-sequence since

$$d(e_i, e_j) = \sqrt{2}, \qquad \forall i, j, \quad i \neq j$$

Since $d(e_i, e_j) = \| e_i - e_j \|$, and

$$\| e_i - e_j \|^2 = \langle e_i - e_j, e_i - e_j \rangle$$
$$= \langle e_i, e_i \rangle + \langle e_j, e_j \rangle - 2\langle e_i, e_j \rangle = 2$$

(It is, however, possible to define a compact subset of l^2; for instance, the set of all points $e = (e_1, e_2, \ldots)$ for which $|e_1| \leqslant \frac{1}{2}, |e_i| \leqslant (\frac{1}{2})^i$ is compact in l^2.)

Now, in general, the weaker the topology, the more likely it is that a set will be compact. (In fact, it is possible to define the so-called indiscrete topology, which is so weak that every set is compact.)

Since we are often dealing with infinite dimensional problems where the valuable property of compactness is not to hand, it is necessary to define weaker forms of compactness. It is then possible to apply these in weak analogues of theorems such as the Weierstrass theorem.

Theorem 5.7. (Banach–Alaoglu theorem). *Let X be a real normed linear space and let β^* be the unit ball in X^*. Then in the weak* topology β^* is compact.*

Let X be a reflexive Banach space then the closed unit ball β in X is weak compact.

Conversely if the closed unit ball β in X is weak compact then X is a reflexive space. (This theorem is proved in reference J3.)

5.14 CONVERGENCE OF OPERATORS

Let $\{T_i\}$ be a sequence in $B(X)$, (see Section 2.14), where X is a Hilbert space. $\{T_i\}$ is said to *converge uniformly* to an operator T if

$$\lim_{i \to \infty} \| T_i - T \| = 0 \tag{5.14.1}$$

$\{T_i\}$ is said to *converge strongly* to an operator T if

$$\lim_{i \to \infty} \| T_i x - T x \| = 0, \qquad \forall x \in X \tag{5.14.2}$$

$\{T_i\}$ is said to *converge weakly* to an operator T if

$$\lim_{i \to \infty} \langle x, T_i x \rangle = \langle x, T x \rangle, \qquad \forall x \in X \tag{5.14.3}$$

If (5.14.1) is satisfied, then so is (5.14.2), since

$$\| T_i x - Tx \| = \|(T_i - T)x\| \leqslant \| T_i - T \| \| x \|$$

(Distinguish carefully in the above expression between the operator norm and the Hilbert space norm.) Thus uniform convergence implies strong convergence.

If (5.14.2) is true, then so is (5.14.3), since

$$\lim_{i \to \infty} \langle x, T_i x \rangle - \langle x, Tx \rangle = 0$$

provided that T is strongly continuous. Thus, uniform convergence $\Rightarrow$ strong convergence $\Rightarrow$ weak convergence.

Later in the book it will be seen that a finite set of ordinary differential equations is representable by a uniformly continuous mapping. Partial differential equations cannot meet the conditions for uniform continuity but can, at best, be represented by a strongly continuous mapping.

5.15 WEAK, STRONG AND UNIFORM CONTINUITY

Let $x(a)$ be a *vector valued function* defined on an abstract set E with values in a Banach space X so that to each element a in E there corresponds a unique element $x(a) \in X$. In case the range of the function is $B(X, Y)$ the function is called an *operator valued function*.

A vector valued function $x(a): \mathbb{R}^1 \to X$, X a Banach space, is:

(i) *weakly continuous* at a_0 if

$$\lim_{a \to a_0} |x^*(x(a) - x(a_0))| = 0, \qquad \forall x^* \in X^*$$

(ii) *strongly continuous* at a_0 if

$$\lim_{a \to a_0} \| x(a) - x(a_0) \| = 0$$

An operator valued function $Q(a): \mathbb{R}^1 \to B(X, Y)$ is:

(i) *continuous in the weak operator topology* (weakly continuous) at a_0 if

$$\lim_{a \to a_0} |y^*\{(Q(a) - Q(a_0))x\}| = 0, \qquad \forall x \in X, \quad y^* \in Y^*$$

(ii) *continuous in the strong operator topology* (strongly continuous) at a_0 if

$$\lim_{a \to a_0} \|(Q(a) - Q(a_0))x\| = 0, \qquad \forall x \in X$$

(iii) *continuous in the uniform operator topology* (uniformly continuous) at a_0 if

$$\lim_{a \to a_0} \| Q(a) - Q(a_0) \| = 0.$$

5.16 EXERCISES

(1) Let $T: X \to Y$ be a mapping of finite rank between Banach spaces. Using a matrix representation for T, show by setting up a coordinate system that, given an element $y \in Y$, the set of elements in X satisfying $T(x) = y$ lie in a finite dimensional subspace of X.

(2) $C[a, b]$ is the space of functions f continuous on the interval $[a, b]$ with norm

$$\| \cdot \| = \sup_{t \in [a, b]} \{ |f(t)| \}$$

Show that $C[a, b]$ is not reflexive.

(3) Prove that the closed unit ball β in a Hilbert space is rotund.

(4) Compare the two properties, uniform convexity and rotundity.

(5) Show by simple manipulation that if p is in the open interval $(1, \infty)$ then the L^p spaces are reflexive.

(6) Let X be a normed linear space and let X^* be its dual space. Let $x^* \in X^*$. Under what conditions is it true that

$$\langle x, x^* \rangle = \| x \| \, \| x^* \|, \qquad x \in X?$$

If X is a Hilbert space, how does this alter the conditions for equality?

(7) Let $\{x_n\}$ be a sequence in a Hilbert space X. Prove that if $\{x_n\}$ converges strongly to an element x in X then $\{x_n\}$ converges weakly to x.

(8) Let X be a real normed linear space. Let $\langle x, x^* \rangle$ be considered as a functional on X^*, i.e. $x \in X$ is regarded as fixed. Let x^* be confined to the unit ball $\beta^* \subset X^*$. Under what conditions can it be guaranteed that $\langle x, x^* \rangle$ will achieve its maximum on β^*?

Axiomatic Representation of Systems

6.1 INTRODUCTION

In this short chapter, system axioms are laid down that so far as possible will be used throughout the later chapters to provide rigour and consistency. The approach used here is based on the papers by Weiss and Kalman (W8, K3).

More generalized abstract approaches to the problem of system description are to be found in Mesarovic and Windeknecht (M3, W12).

6.2 THE AXIOMS

Let I be the real line $\mathbb{R}^1$. Let U, X, Y be normed linear spaces of functions on I. Let Ω be a non-empty subset of U. Let $\phi \colon X \times I \times \Omega \times I \to X$ be a mapping, linear in the first term and locally linear (see Section 1.12) in the third term. Let $\eta \colon X \times I \to Y$ be a linear mapping.

A linear system, denoted by Σ, can then be represented

$$\Sigma = \{I, U, \Omega, X, Y, \phi, \eta\}$$

U, X, Y will be referred to as the *input, state* and *output spaces* respectively, Ω is referred to as the set of *admissible inputs*. ϕ will be referred to as the *state transition mapping* and η as the *output mapping*. ϕ has the property

$$\phi\big(x(t_0), t_0, u(t), t_1\big) = x(t_1), \qquad t \in [t_0, t_1]$$

The linearity of the state transition mapping allows the following useful decomposition to be made,

$$\phi\big(x(t_0), t_0, u(t), t_1\big) = \phi\big(0, t_0, u(t), t_1\big) + \phi\big(x(t_0), t_0, 0, t_1\big)$$

The system Σ will be defined to be a *dynamic system* if it satisfies the two axioms below.

Axiom 1. (Semi-group property)

$$\phi(x(t_0), t_0, u(t), t_2) = \phi(\phi(x(t_0), t_0, u(t), t_1), t_1, u(t), t_2).$$

Axiom 2. (Causality condition)

$$\phi(x(t_0), t_0, u_1(t), t_1) = \phi(x(t_0), t_0, u_2(t), t_1), \qquad t_0, t_1 \in I, \qquad t_1 > t_0$$

$$u_1(t) = u_2(t), \qquad t \in [t_0, t_1]$$

$$u_1(t) \neq u_2(t), \qquad \text{otherwise}$$

The system Σ is defined to be *time-invariant* if the mappings ϕ and η are time invariant, satisfying Axiom 3.

Axiom 3. (Time invariance)

$$\phi(x(t_0), t_0, u(t), t_1) = \phi(x(t_0), t_0 + T, u(t), t_1 + T),$$

$$\eta(x(t_0), t_0) = \eta(x(t_0), t_0 + T), \qquad t_0, t_1, T \in I$$

In the time invariant case, the mappings ϕ and η can be simplified as shown below. Note that for convenience these simplified mappings continue to be denoted by the symbols ϕ, η:

$$\phi = \phi(x(t_0), u(t), t_1 - t_0), \qquad x \in X, \quad t_1, t_0 \in I$$

$$\eta = \eta(x), \qquad x \in X$$

Axiom 4. (Alternative to axiom 3, i.e. in case the mappings ϕ, η are not time invariant.) *The mappings ϕ, η are continuous functions of time.*

Axiom 5. *The space U is a space of piecewise-continuous functions on I.*

Axiom 3 ensures that the system is time invariant while Axiom 4 ensures that, although the system may be time varying, it will satisfy conditions for the existence and uniqueness of solutions. In fact, Axiom 4 is sharper than is needed for this purpose.

Axiom 5 ensures existence and uniqueness of a solution in the presence of inputs. In practice, every physically realizable input is measurable and Axiom 5 causes no restriction to the types of control that can be considered.

The axioms are discussed more fully in Section 6.4.

Summary

The most general representation of a linear system is that given by

$$\Sigma = \{1, U, \Omega, X, Y, \phi, \eta\}$$

with the requirement that the spaces and mappings are linear. The system Σ is called a *linear dynamic system* provided that it satisfies Axioms 1, 2, 4 and 5. In the special case that Axiom 3 is also satisfied the system is referred to as a *linear time invariant dynamic* system.

Definition. *Let Σ be a linear dynamic system satisfying the requirement $X = \mathbb{R}^n$. Then Σ is called a linear dynamic system with finite dimensional state space.* Since this work is concerned exclusively with linear dynamic systems, Σ will be referred to simply as a *finite dimensional system*.

Note

A dynamic system such as we have defined with $I = \mathbb{R}^1$ is referred to as a *continuous time system*. If instead we had defined I to be a set of isolated points in $\mathbb{R}^1$ then Σ would in this case have been a *discrete time system*. The most usual case is where the isolated points are spaced at equal intervals in $\mathbb{R}^1$. Normalizing this interval to unity, the set of isolated points becomes Z and the system is called a *sampled data system*. Discrete time system are not considered specifically in this book. References C1, P7 can be consulted on the detail of formulation of discrete time models within the above framework.

6.3 RELATION BETWEEN THE AXIOMATIC REPRESENTATION AND THE REPRESENTATION AS A FINITE SET OF DIFFERENTIAL EQUATIONS

Let Σ be a continuous time finite dimensional system for which

$$X = \mathbb{R}^n, \qquad Y = \mathbb{R}^m$$

Then Σ can be characterized by the equation

$$\left. \begin{aligned} \dot{x}(t) &= f(x, u, t) = A(t)x(t) + B(t)u(t) \\ y(t) &= C(t)x(t) \end{aligned} \right\} \qquad (6.3.1)$$

A is an $n \times n$ matrix; B is an $n \times r$ matrix; C is an $m \times n$ matrix.

The axioms of Section 6.2 ensure that Σ represents a linear dynamic system possessing a unique solution. Thus, not only is the system Σ represented by equation (6.3.1) but the axioms of Section 6.2 ensure that equation (6.3.1) is well posed, i.e. that it possesses a solution, the solution is unique and the solution is a continuous function of the initial conditions.

The solution of equation (6.3.1) can then be written in terms of the transition matrix, Φ,

$$\left.\begin{aligned} x(t) &= \Phi(t, t_0)x_0 + \int_{t_0}^{t} \Phi(t, \tau)B(\tau)u(\tau)\, d\tau \\ y(t) &= C(t)x(t) \end{aligned}\right\} \tag{6.3.2}$$

In the time invariant case, equation (6.3.1) reduces to

$$\begin{aligned} \dot{x}(t) &= Ax(t) + Bu(t) \\ y(t) &= Cx(t) \end{aligned} \tag{6.3.3}$$

with solution

$$\begin{aligned} x(t) &= \Phi(t - t_0)x_0 + \int_{t_0}^{t} \Phi(t - \tau)Bu(\tau)\, d\tau \\ y(t) &= Cx(t) \end{aligned} \tag{6.3.4}$$

The linear mappings ϕ, η can be identified within the sets of equations (6.3.2) or (6.3.4) and this aspect is pursued below. (The term ΦB is sometimes denoted by Ψ.)

6.4 VISUALIZATION OF THE CONCEPTS OF THIS CHAPTER

This section is designed to help geometric visualization of the meaning of the axioms and to point out the inter-relation between the axiomatic formulation and the differential equation formulation.

Let a linear system $\Sigma = \{I, U, \Omega, X, Y, \phi, \eta\}$ be formulated so that $I = \mathbb{R}^1$, $U = (L^p)^r$. Ω is the set of piecewise continuous functions in U, $X = \mathbb{R}^n$, $Y = \mathbb{R}^m$ with ϕ, η being time invariant mappings satisfying Axioms 1, 2, 3, 5.

Consider the vector-matrix differential equation

$$\begin{aligned} \dot{x} &= Ax + Bu, \qquad x(0) = x_0 \\ y &= Cx \end{aligned} \tag{6.4.1}$$

where A, B, C are constant matrices of dimension $n \times n$, $n \times r$, $m \times n$ respectively.

Under the conditions assumed, we know from differential equation theory that the equation (6.4.1) is well posed, i.e. that the equation possesses a solution, that the solution is unique and that it depends continuously on the initial condition vector x_0.

Under such conditions, the "solution" of (6.4.1),

$$\left.\begin{aligned} x(t) &= e^{A(t-0)}x_0 + \int_0^t e^{A(t-\tau)}Bu(\tau)\, d\tau \\ y(t) &= Cx(t) \end{aligned}\right\} \tag{6.4.2}$$

can be considered to be an exact equivalent of the original equation (6.4.1). We shall use equation (6.4.2) to illustrate the meaning of the axioms. Figure 6.1 shows the relation between equation (6.4.2) and the mappings ϕ, η. It also shows the decomposition of the mapping ϕ as given in Section 6.2.

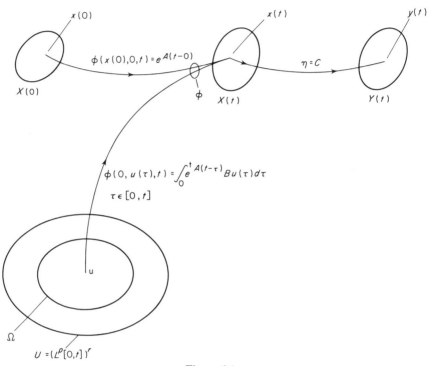

Figure 6.1.

Axiom 1: Semi-group property

Axiom 1 is obviously a necessary condition if the system Σ is to have a unique solution. To see this assume the axiom does not hold then we would have

$$x(t_2) = \phi(x(t_0), t_0, u(t), t_2) \neq \phi(\phi(x(t_0), t_0, u(t), t_1), t_1, u(t), t_2)$$

But the right-hand term is equal to

$$\phi(x(t_1), t_1, u(t), t_2) = x(t_2)$$

a contradiction.

Figure 6.2 may be helpful in visualizing the concept for the simple case where $u = 0$

Consider computing the time solution of equation (6.4.2) with $x(0) = x_0$, $u(t) = 0$. Let $t_1, t_2 \in \mathbb{R}^1, 0 < t_1 < t_2$ then

$$x(t_1) = e^{A(t_1 - 0)}x_0 \qquad (6.4.3)$$

$$x(t_2) = e^{A(t_2 - 0)}x_0 \qquad (6.4.4)$$

But also

$$x(t_2) = e^{A(t_2 - t_1)}x(t_1) \qquad (6.4.5)$$

(These three equations are illustrated in Figure 6.2.)

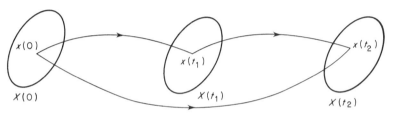

Figure 6.2.

If uniqueness is to hold then $x(t_2)$ must be the same in both equations (6.4.4), (6.4.5) above, hence

$$e^{A(t_2 - 0)}x_0 = e^{A(t_2 - t_1)}x(t_1) = e^{A(t_2 - t_1)}e^{A(t_1 - 0)}x_0$$

which is precisely the semi-group property of Axiom 1.

Axiom 2: Causality

Causality is closely linked with the question of physical realizability. Intuitively, a causal system is affected by the past but not by the future. This means that a system has to behave asymmetrically with respect to time in order to be causal and hence physically realizable.

Causality can be considered to be linked with the irreversibility of time and with thermodynamic laws. In the case of equation (6.4.2) it is clear that x_0 summarizes the effect of past inputs over the interval $(-\infty, 0]$, while the convolution integral determines the effects of the input $u(\tau)$ over the interval $[0, t]$. Clearly the equation (6.4.2) is causal since $u(\tau), \tau > t$ can have no effect.

Axiom 3: Time invariance

Clearly the differential equation (6.4.1) is time-invariant since the matrices A, B, C are time invariant. However, we can illustrate directly how Axiom 3 applies in relation to equation (6.4.2).

According to this axiom we must have, for every $T \in I$

$$\Phi(t_1 - t_0)x_0 + \int_{t_0}^{t_1} \Phi(t_1 - \tau)Bu(\tau)\, d\tau = \Phi(t_1 + T - (t_0 + T))$$

$$+ \int_{t_0 + T}^{t_1 + T} \Phi(t_1 + T - \tau)Bu(\tau)\, d\tau$$

The right-hand side is equal to

$$\Phi(t_1 - t_0)x_0 + \int_{r_0}^{r_1} \Phi(r_1 - \tau)Bu(\tau)\, d\tau \qquad (\text{where } r_i = t_i + T)$$

$$= \Phi(t_1 - t_0)x_0 + \int_{t_0}^{t_1} \Phi(t_1 - \tau)Bu(\tau)\, d\tau$$

(since $r_1 - r_0 = t_1 - t_0$ and Φ is a time invariant mapping dependent only on the interval $t_1 - t_0$).

Notice several important points. The differential equation for a time varying system has in general matrices $A(t)$, $B(t)$ of time varying coefficients. Under this condition the transition matrix is a function of two variables and is denoted $\Phi(t_1, t_0)$ to indicate this. $B(t)$ appears in the convolution integral and it is clear that the presence of either $B(t)$ or $\Phi(t_1, t_0)$ in the solution ensures that the time invariance axiom cannot be satisfied.

If for *some* value of T, Axiom 3 is satisfied, or equivalently if for some value of T,

$$\Phi(t_1, t_0)x(t_0) + \int_{t_0}^{t_1} \Phi(t_1 - \tau, t_0)B(\tau)u(\tau)\, d\tau$$

$$= \Phi(t_1 + T, t_0 + T)x(t_0) + \int_{t_0 + T}^{t_1 + T} \Phi(t_1 + T - \tau, t_0 + T)B(\tau)u(\tau)\, d\tau$$

then the system Σ is *periodic*, and the period of the system is the smallest value of T for which the axioms can be satisfied.

Axiom 4: Alternative to Axiom 3

This expresses the fact that some restriction must be imposed on the types of functions permitted in the $A(t)$, $B(t)$ matrices if the system is to be well posed.

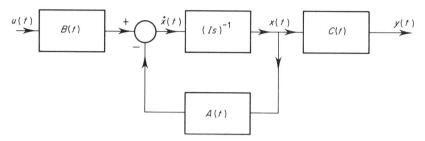

Figure 6.3. System described by $\dot{x}(t) = A(t)x(t) + B(t)u(t)$, $y(t) = C(t)x(t)$

Axiom 5

This is similar to Axiom 4 in that it restricts the functions $u(t)$ that can be input to the system thereby ensuring that the system is well posed.

Visualization of the equation (6.4.1) may be assisted by Figure 6.3. Figure 6.4 shows the block diagram of a general feedback system with feedback controller $K(t)$.

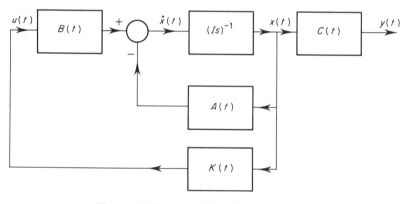

Figure 6.4. System with feedback control

6.5 SYSTEM REALIZATION

Consider the following situation. A system, known to be linear, takes in a number of input signals $u_1(t), \ldots, u_r(t)$ and gives out a number of output signals $y_1(t), \ldots, y_m(t)$. Nothing more is known about the system which can be thought of as the celebrated black box. In particular nothing whatever is known *a priori* about any state space X that may or may not exist. The question of *realization* is then: Can the system just discussed be described as in Section 6.2? More precisely, can a system described only by input–output data be described in axiomatic or state space form? If the answer is yes, is such a realization unique and can the realization be performed by feasible algorithms operating on the input–output data?

Let the spaces U, X, Y be defined as earlier in Section 6.2. Then a particular input–output relation must be expressed by the composition $\eta \circ \phi: U \to Y$ or equivalently by the vector-matrix differential equation

$$\dot{x} = Ax + Bu, \qquad y = Cx \qquad (6.5.1)$$

Assume that such representations exist and that the matrix A has distinct eigenvalues. Put $x = Ez$ where (see Section 6.7) E is the modal matrix of A. The differential equation (6.5.1) then becomes

$$E\dot{z} = AEz + Bu, \qquad y = CEz$$

or

$$\dot{z} = (E^{-1}AE)z + E^{-1}Bu, \qquad y = CEz$$

This shows that the same input–output relation can be represented by another pair of mappings $\eta' \circ \phi': U \to Y$ through an intermediate state space Z. Thus, system realizations are not unique.

For a particular input–output relation it is not even true that the dimension of the state space is uniquely determined. However, there does exist a dimension n that is the lowest possible. A realization utilizing a state space of least possible dimension is known as a *minimal realization*.

The topic of realization is usually studied using system impulse responses to make precise the input–output relation that it is desired to realize and proceeding to canonical vector-matrix realizations. The question of realization is closely linked to the properties of system controllability and observability. These properties are considered in the next chapter.

6.6 THE TRANSITION MATRIX Φ AND SOME OF ITS PROPERTIES

Definition

The transition matrix $\Phi(t, t_0)$

is defined by

$$\Phi(t, t_0)x_0 = \phi(x_0, t_0, 0, t)$$

Properties of Φ

Composition

$$\Phi(t, t_0) = \Phi(t, t_1).\Phi(t_1, t_0)$$

$$\Phi(t_1, t_0) = \Phi(t_1, 0).\Phi(0, t_0) = \Phi(t_1, 0).\Phi^{-1}(t_0, 0)$$

Differentiation

$$\frac{d\Phi(t, t_0)}{dt} = A(t)\Phi(t, t_0)$$

Characterization

If A is constant then

$$\Phi(t, t_0) = I + A(t - t_0) + \frac{A^2(t - t_0)^2}{2!} + \ldots = e^{A(t - t_0)}$$

$$e^{At}e^{A\tau} = e^{A(t + \tau)}$$

but note that

$$e^A e^B \neq e^{(A + B)} \quad \text{unless} \quad AB = BA$$

and thus in general

$$\Phi(t, t_0) \neq \exp\left(\int_{t_0}^{t} A(\sigma)\, d\sigma\right)$$

The Baker–Campbell–Hausdorff theorem gives formulae for combining the transition matrices of non-commutative square matrices (see reference B11).

6.7 CALCULATION OF THE TRANSITION MATRIX Φ FOR TIME INVARIANT SYSTEMS

In numerical computation it is frequently necessary to calculate $\Phi(t)$ given an $n \times n$ matrix A of constant coefficients. There are several methods available.

(1) Use the series for e^{AT} given above, for a small fixed value of T. The series is always convergent and quickly convergent for T very small. Given $\Phi(T)$ we can quickly calculate the solution of the equation $\dot{x}(t) = Ax(t)$, $x(0) = x_0$ at discrete values of T by using the equation

$$x(T) = \Phi(T)x(0), \quad x(2T) = \Phi(T)x(T)$$

and in general

$$x((k + 1)T) = \Phi(T)x(kT)$$

(2) We can easily show that

$$\Phi(t) = \mathcal{L}^{-1}((sI - A)^{-1})$$

where $\mathcal{L}^{-1}$ indicates inverse Laplace transformation and s is the complex variable associated with $\mathcal{L}$. This method is useful when we wish to obtain an analytic expression for $\Phi(t)$ but it may be difficult to invert the transforms on occasions.

(3) If fortuitously A were a diagonal matrix,

$$A = \begin{pmatrix} a_1 & & & \\ & a_2 & & 0 \\ & & \ddots & \\ 0 & & & a_n \end{pmatrix} \quad \text{then} \quad \Phi(t) = \begin{pmatrix} e^{a_1 t} & & & \\ & e^{a_2 t} & & 0 \\ & & \ddots & \\ 0 & & & e^{a_n t} \end{pmatrix}$$

Let A be a matrix with distinct eigenvalues $\lambda_1, \ldots, \lambda_n$ and eigenvectors $e_1, \ldots, e_n$. Let $E = (e_1 | \ldots | e_n)$ be the so called *modal matrix* of A. E^{-1} always exists since the e_i are linearly independent. Then it can easily be shown that $E^{-1}AE = \Lambda$ where

$$\Lambda = \begin{pmatrix} \lambda_1 & & 0 \\ & \ddots & \\ 0 & & \lambda_n \end{pmatrix}$$

$E^{-1}(\cdot)E$ represents a transformation to new coordinates. (The new coordinate axes are in fact the eigenvectors.) Now

$$e^{\Lambda t} = \begin{pmatrix} e^{\lambda_1 t} & & 0 \\ & \ddots & \\ 0 & & e^{\lambda_n t} \end{pmatrix}$$

Transforming back to the original coordinates by the operation $E(\cdot)E^{-1}$ yields

$$\Phi(t) = Ee^{(E^{-1}AE)t}E^{-1}$$

$$= E \begin{pmatrix} e^{\lambda_1 t} & & 0 \\ & \ddots & \\ 0 & & e^{\lambda_n t} \end{pmatrix} E^{-1}$$

(4) By Sylvester's theorem, provided that A has distinct eigenvalues,

$$\Phi(t) = e^{At} = \sum_{i=1}^{n} e^{\lambda_i t} F(\lambda_i)$$

where n is the order of the matrix A and

$$F(\lambda_i) = \prod_{\substack{j=1 \\ j \neq i}}^{n} \left(\frac{A - \lambda_j I}{\lambda_i - \lambda_j} \right)$$

6.8 EXERCISES

(1) Prove that the set $\{e^{AT}\}$ forms a semi-group of linear transformations. Does the set in fact form a group?

(2) The operator $Tx(t)$ defined by

$$Tx(t) = \int_{-\infty}^{t} h(t - \tau)x(\tau) \, d\tau$$

is important in dynamic systems modelling and is known as a *Volterra operator*. Show that the Volterra operator is time invariant and causal. If $T \in B(X, Y)$, x, y Hilbert spaces, then the *adjoint* T^* of T is defined by

$$\langle x, T^*y \rangle = \langle Tx, y \rangle, \qquad \forall x \in X, \quad \forall y \in Y$$

Show that the adjoint of T is also an integral operator but that it is no longer causal.

The convolution integral can be defined

$$Sx(t) = \int_{-\infty}^{\infty} h(t - \tau)x(\tau) \, d\tau$$

Is the operator S causal?

(3) Show that the system described by $\dot{x}(t) = u(t + T)$, $T > 0$, does not satisfy the axioms for a dynamic system.

(4) Let $\dot{x} = Ax + Bu$, A, $n \times n$ and B, $n \times r$, let $y = Cx$. Put $x = Dz$ where D is any non-singular $n \times n$ matrix. Put the original equations in the form $z = A'z + B'u$, $y = C'z$. Satisfy yourself that A and A' have the same eigenvalues, thus proving that a choice of state variables exists.

(5) A matrix N is defined to be *nilpotent* if $N^m = 0$ for some integer $m > 1$. Suppose a matrix A can be decomposed so that $A = S + N$, where S is diagonal and N is nilpotent. Show how the decomposition can be exploited to allow rapid calculation of the matrix exponential e^A.

(6) Let A be a square matrix with distinct eigenvalues. Let E be the modal matrix $(e_1|e_2|\ldots e_n)$ where e_i are the eigenvectors of A. Prove that

$$E^{-1}AE = \begin{pmatrix} \lambda_1 & & 0 \\ & \ddots & \\ 0 & & \lambda_n \end{pmatrix}$$

and hence confirm that

$$e^A = Ee^{E^{-1}AE}E^{-1}$$

(7) Let A be a symmetry matrix. Show that the eigenvectors of A are orthogonal to each other.

(8) A system is described by the equations

$$\dot{x}_1 = x_1 + 4x_2 + u, \qquad \dot{x}_2 = 3x_1 + 2x_2$$

Determine the transition matrix and write down the general solution.

(9) A system is described by the equation

$$\ddot{y} + 3\dot{y} + 2y = u, \qquad y(0) = \dot{y}(0) = 0$$
$$u(t) = 1, \qquad 0 \leqslant t < 1$$
$$u(t) = 0, \qquad t \geqslant 1$$

Calculate the transition matrix and determine $y(2)$.

CHAPTER 7

Stability, Controllability and Observability

7.1 INTRODUCTION

Stability, controllability and observability are important structural properties of dynamic systems. In general, powerful techniques of functional analysis are not required for these topics, although Freeman (F6, F7) has produced rather general stability criteria using fixed point theorems applied to spaces whose elements are integral transforms. The stability results given here are standard in control theory except for the part of Section 7.2 dealing with Lyapunov's method.

Controllability is concerned with establishing that the set of controls Δ that can accomplish a stated task is non-empty. (Later, optimal control theory will be concerned with determining the element in Δ that is best in some sense.) The controllability theorems given here are basically those of Kalman (K1, K4), who was the first to define controllability.

Observability is a dual property of controllability and most results can be obtained by parallel arguments to those used in the controllability sections. Accordingly, observability is covered very briefly.

In relation to this chapter it is emphasized that the results are concerned exclusively with systems with finite dimensional state space. For systems with infinite dimensional state spaces the topics are more complex. As an example, there is no universally accepted definition of controllability for such systems. Chapter 10 takes up this point.

7.2 STABILITY

7.2.1 Discussion

There are several possible definitions of stability. Broadly speaking, for a stable system,

$$\phi(x_0, t_0, u, t) \simeq \phi(x_0 + \delta x_0, t_0, u, t), \qquad \forall t > t_0$$

where δx_0 is a small perturbation in the initial state. A more stringent stability requirement is that

$$\lim_{t \to \infty} (\phi(x_0 + \delta x_0, t_0, u, t)) = \lim_{t \to \infty} (\phi(x_0, t, u, t))$$

with the rate of convergence perhaps being subject to special requirements.

Finally, attention is devoted to the effect of the input, $u(t)$, where $u(t)$ is assumed to belong to a particular class of functions.

7.2.2 Definitions of stability

(1) A system Σ will be defined to be *stable* (often called *stable in the sense of Lyapunov*) if given x_0, t_0 there exists $k(t_0, x_0) \in K$ such that

$$\| \phi(x_0, t_0, 0, t) \| \leqslant k(t_0, x_0), \qquad \forall t \geqslant t_0$$

(2) A system Σ will be defined to be *asymptotically stable* if

$$\| \phi(x_0, t_0, 0, t) \| \to 0 \quad \text{as} \quad t \to \infty, \qquad \forall x_0 \in X, \quad \forall t_0 \in I$$

Clearly $(2) \Rightarrow (1)$.

(3) A system Σ is defined to be *bounded-input-bounded-output stable* if for every scalar $k > 0$ there exists M such that

$$\| u(t) \| \leqslant M \Rightarrow \| \eta \phi(0, t_0, u, t) \| \leqslant k, \qquad \forall t_0, \quad \forall t \geqslant t_0$$

with u defined on $[t_0, t]$.

(4) A system Σ is defined to be *bounded-input-bounded-state stable* if for all u on $[t_0, \infty)$ satisfying $\| u \| \leqslant k < \infty$ there exists a real number M such that

$$\| \phi(x_0, t_0, u, t) \| \leqslant M(x_0, t_0, k), \qquad \forall t \geqslant t_0, \quad \forall t_0 \in I$$

Remark

Since the output mapping η is continuous, $(4) \Rightarrow (3)$ while $(3) \Rightarrow (4)$ if and only if the system Σ is observable (see Section 7.3.4).

7.2.3 Stability for time invariant systems

Theorem 7.1. *Let Σ be a time invariant system that can be described by the equation*

$$\dot{x}(t) = Ax(t), \qquad x(t) \in X \times I, \quad X = \mathbb{R}^n \tag{7.2.1}$$

Then

(i) Σ *is stable if each eigenvalue* λ_i *of A satisfies* $\mathscr{R}(\lambda_i) \leqslant 0$.
(ii) Σ *is asymptotically stable if each eigenvalue* λ_i *of A satisfies the relation* $\mathscr{R}(\lambda_i) < 0$.

Proof. Assume that A has distinct eigenvalues. Equation (7.2.1) can be transformed by a change of coordinates into the equivalent representation:

$$\dot{q}(t) = \Lambda q(t)$$

where Λ is a diagonal matrix of the eigenvalues of A. Hence,

$$q_i(t) = e^{\lambda_i t} q_i(t_0)$$

Let $k = \| q(t_0) \|$ then

$$\lambda_i \leqslant 0 \, (i = 1, \ldots, n) \Rightarrow \| q(t) \| \leqslant k, \qquad (t \geqslant t_0)$$

and the system Σ is stable.

$$\lambda_i < 0 \, (i = 1, \ldots, n) \Rightarrow \| q(t) \| \to 0 \quad \text{as } t \to \infty$$

and the system is asymptotically stable.

(If the matrix A has repeated eigenvalues we can proceed as follows. Write the solution of (7.2.1) in the general form

$$x(t) = P_1(t) e^{\lambda_1(t)} + \ldots + P_k e^{\lambda_k(t)}$$

where each P_i is a polynomial in t of order equal to the multiplicity of the ith eigenvalue. It is clear that if all the λ_i have strictly negative real part then the system is asymptotically stable to the origin. The case where repeated eigenvalues have zero real part needs further consideration but this point will not be pursued here.)

7.2.4 Lyapunov's second method interpreted for linear time invariant systems

Let Σ be a system described by the equation

$$\dot{x}(t) = A x(t), \qquad x \in \mathbb{R}^n, \qquad \det(A) \neq 0$$

The Lyapunov second (or direct) method asserts that if there exists a function $V: \mathbb{R}^n \to \mathbb{R}^1$ that is positive definite and for which $\dot{V}: \mathbb{R}^n \to \mathbb{R}^1$ evaluated along the system trajectories is negative definite then the system Σ is asymptotically stable to the origin.

(The condition $\det A \neq 0$ is not significant but is inserted to ensure that the special case where there is a zero eigenvalue is excluded.)

The scalar valued function V is called a Lyapunov function and it is a generalization of the total energy in the system. Notice that the stability question is answered without any solution of the differential equations being required—this is the attraction of the method.

For linear systems, the Lyapunov direct method is little used since there are many alternative methods available. (In particular, the Hurwitz test, see reference Z1). It is, however, instructive to interpret the method for linear systems as follows.

Consider again the system Σ and the equation (7.2.1). Let F be a positive definite matrix then the inner product $V = \langle x, Fx \rangle$ is positive definite and is therefore a possible Lyapunov function. To satisfy the requirement we must have $\mathrm{d}(\langle x, Fx \rangle)/\mathrm{d}t$ negative definite.

$$\frac{\mathrm{d}(\langle x, Fx \rangle)}{\mathrm{d}t} = \langle \dot{x}, Fx \rangle + \langle x, F\dot{x} \rangle$$
$$= \langle Ax, Fx \rangle + \langle x, FAx \rangle$$
$$= \langle x, (A^T F + FA)x \rangle$$
$$= \langle x, Gx \rangle$$

where $G = A^T F + FA$.

If G is negative definite then $\dot{V}$ will be negative definite, V will be a Lyapunov function and, invoking the relevant Lyapunov theorem, the system Σ will be stable to the origin.

We show that F will be positive definite if G is negative definite and if all the eigenvalues of the matrix A have negative real parts. Thus, for linear time invariant systems, the Lyapunov method can be nicely linked with other methods of stability investigation.

Assume that all eigenvalues of A have negative real parts. Let

$$\dot{Q}(t) = A^T Q(t) + Q(t)A \tag{7.2.2}$$

where $Q(t)$ is a square matrix whose elements are continuous functions of time with $Q(0) = -G$. The solution of equation (7.2.2) is

$$Q(t) = -e^{A^T t} G\, e^{At}$$

(Differentiate this equation to confirm that it satisfies equation (7.2.2)). Thus $\lim_{t \to \infty} Q(t) = 0$, since the eigenvalues of A have negative real part. Then

$$\int_0^\infty \dot{Q}(t)\, \mathrm{d}t = 0 - Q(0) = G$$
$$= A^T \int_0^\infty Q(t)\, \mathrm{d}t + \int_0^\infty Q(t)\, \mathrm{d}t \,.\, A.$$

Recall that $G = A^T F + FA$, hence

$$F = \int_0^\infty Q(t)\, \mathrm{d}t = -\int_0^\infty e^{A^T t} G e^{At}\, \mathrm{d}t$$

$$\langle x, Fx \rangle = - \left\langle x, \int_0^\infty e^{A^T t} G e^{At} \, dt \cdot x \right\rangle$$

The right-hand term becomes

$$-\int_0^\infty \langle e^{At} x, G e^{At} x \rangle \, dt > 0, \qquad \text{any constant } x \neq 0,$$

since the matrix G is negative definite. Thus F is positive definite. (This proof follows reference R5.)

7.2.5 Geometric interpretation of the Lyapunov second method

Let Σ be a linear time invariant system with state space $X = \mathbb{R}^2$. Let $V: X \to \mathbb{R}^1$ be positive definite. Then putting different values of k in the equation $V(x_1, x_2) = k$ results in contours of constant V in the x_1, x_2 plane (see Fig. 7.1). Now to say that $\dot{V}$ is negative definite is the same as to say that $\langle \dot{x}, \text{grad } V \rangle < 0$ everywhere in the x_1, x_2 plane. Choose an arbitrary point (z_1, z_2) in the x_1, x_2 plane. The system motion through the point

$$z = \begin{pmatrix} z_1 \\ z_2 \end{pmatrix}$$

is described instantaneously by the vector $\dot{z}$.

If $\langle \dot{z}, \text{grad } V \rangle < 0$ then clearly the component of the system motion normal to the V contour must be given by $-C \text{ grad } V$ for some positive constant C. More formally, the projection of the system motion $\dot{z}$ on to grad V at z is given by

$$l = \left\langle \dot{z}, \frac{\text{grad } V}{\|\text{grad } V\|} \right\rangle \frac{\text{grad } V}{\|\text{grad } V\|} \tag{7.2.3}$$

This equation can be visualized with the help of Fig. 7.2.

The geometric interpretation is that if V is negative definite then $\langle \dot{x}, \text{grad } V \rangle < 0$ everywhere in the plane and that the system state must move asymptotically to the origin.

Finally, since $\dot{x} = Ax$, it is clear that we can check on stability by investigating the sign properties of $\langle Ax, \text{grad } V \rangle$ without needing to solve the differential equation.

7.2.6 Stability theorems for time varying systems

Theorem 7.2. *Let Σ be a continuous time finite state system, then the condition for stability reduces to the following: The system Σ is stable if given x_0 there*

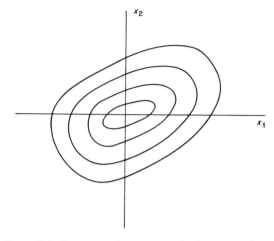

Figure 7.1. Contours of constant V in the x_1, x_2 plane.

exists $M(t_0) < \infty$ such that

$$\|\phi(x_0, t_0, 0, t)\| \leqslant M(t_0)\|x_0\|, \qquad \forall t_0 \in I, \quad \forall t \geqslant t_0.$$

Proof. (Refer back to Section 7.2.2.)

$$k(t_0, x_0) = M(t_0)\|x_0\|$$

by linearity of Σ.

Theorem 7.3. *The system Σ is stable if and only if there exists a finite number N such that*

$$\|\Phi(t, 0)\| \leqslant N, \qquad \forall t \geqslant 0.$$

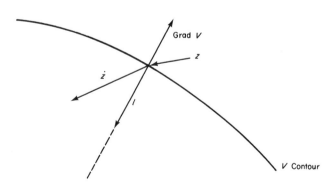

Figure 7.2. Interpretation of equation (7.2.3).

Proof. Assume the bound N exists:

$$\|\phi(x_0, t_0, 0, t)\| = \|\Phi(t, t_0)x_0\|$$
$$= \|\Phi(t, 0)\Phi(0, t_0)x_0\|$$
$$\leqslant \|\Phi(t, 0)\| \|\Phi(0, t_0)\| \|x_0\|$$
$$\leqslant N\|\Phi(0, t_0)\| \|x_0\|.$$

Let $M(t_0) = N\|\Phi(0, t_0)\|$, then by Theorem 7.2 the system is stable.

Assume now that no bound N exists for Φ then the mapping ϕ is unbounded on $[0, \infty)$ and no finite $M(t_0)$ can exist to satisfy Theorem 7.2.

Theorem 7.4. *Let Σ be a system representable by the equation*

$$\dot{x}(t) = A(t)x(t), \qquad x \in X = \mathbb{R}^n.$$

Assume that for all $t \in I$ the eigenvalues of $A(t)$ satisfy the requirement that

$$\mathscr{R}(\lambda_i) < 0, \qquad i = 1, \ldots, n$$

then the system is not necessarily stable.

Proof. (By counter-example due to Markus.) Let

$$A(t) = \begin{pmatrix} a\cos^2(t) - 1 & 1 - a\sin(t)\cos(t) \\ -1 - a\sin(t)\cos(t) & a\sin^2(t) - 1 \end{pmatrix}$$

Then

$$\Phi(t, 0) = \begin{pmatrix} e^{(a-1)t}\cos t & e^{-t}\sin t \\ -e^{(a-1)t}\sin t & e^{-t}\cos t \end{pmatrix}$$

Now let $1 < a < 2$, then

$$x(t) = \Phi(t, 0)x(0)$$

and the system is clearly unstable. However, the eigenvalues of $A(t)$ are independent of t and are given by

$$\lambda = -\alpha \pm \sqrt{(\alpha^2 - 2\alpha)}$$

(where $\alpha = (2 - a)/2$ so that $0 < \alpha < \frac{1}{2}$) so that the metric $A(t)$ has a complex pair of eigenvalues in the left half plane, satisfying $\mathscr{R}(\lambda_i) < 0$ for $i = 1, 2$.

Theorem 7.5. *The system Σ is asymptotically stable if and only if*

$$\|\Phi(t, 0)\| \to 0 \qquad as \ t \to \infty.$$

Proof. From the continuity of ϕ, $\|\Phi(0, t_0)\|$ is finite if t_0 is finite, hence

$\|\Phi(t_0, 0)\|$ is finite. Let $\|\Phi(t_0, 0)\| = k$,

$$\|\Phi(t, 0)\| = \|\Phi(t, t_0)\Phi(t_0, 0)\|$$
$$\leqslant \|\Phi(t, t_0)\| \, \|\Phi(t_0, 0)\|$$
$$= k\|\Phi(t, t_0)\|$$

$$\|\Phi(t, 0)\| \leqslant k \frac{\|\phi(x_0, t_0, 0, t)\|}{\|x_0\|}$$

Hence if the system is asymptotically stable,

$$\|\Phi(t, 0)\| \to 0 \qquad \text{as } t \to \infty.$$

Now assume that $\|\Phi(t, 0)\| \to 0$ as $t \to \infty$,

$$\phi(x_0, t_0, 0, t) = \Phi(t, t_0)x_0$$
$$= \Phi(t, 0)\Phi(0, t_0)x_0$$

$$\|\phi(x_0, t_0, 0, t)\| \leqslant \|\Phi(t, 0)\| \, \|\Phi(0, t_0)\| \, \|x_0\|$$

$\|\Phi(0,t_0)\|$ is finite for finite $t_0 \in I$, hence $\|\Phi(t, 0)\| \to 0$ as $t \to \infty$ implies asymptotic stability of the system Σ.

Theorem 7.6. *The system Σ is bounded-input-bounded-state stable if and only if*

(i) *There exists a real constant N such that*

$$\|\Phi(t, 0)\| \leqslant N, \qquad \forall t \geqslant 0 \quad \text{and}$$

(ii) *There exists a real constant M such that*

$$\int_0^t \|\Phi(t, \tau)B(\tau)\| \, d\tau \leqslant M, \qquad \forall t \geqslant 0.$$

Proof. Let t_0, x_0 be arbitrary and let u satisfy $\sup_{t \geqslant t_0} \|u(t)\| = k$. Then

$$\phi(x_0, t_0, u, t) = \Phi(t, t_0)x_0 + \int_{t_0}^t \Phi(t, \tau)B(\tau)u(\tau) \, d\tau \qquad (7.2.4)$$

$\Phi(t, t_0)x_0$ is bounded from condition (i) and from Theorem 7.3

$$\int_{t_0}^t \|\Phi(t, \tau)B(\tau)u(\tau)\| \, d\tau \leqslant kM$$

conditions (i), (ii) being necessary and sufficient for boundedness of the right-hand side of equation (7.2.4).

Equivalently (i) (ii) are necessary and sufficient conditions for bounded-input bounded-state stability of the system Σ.

7.3 CONTROLLABILITY AND OBSERVABILITY

7.3.1 Preliminaries and definitions

Suppose it is required to design a control system to achieve a particular given objective. For instance, to guide a rocket to rendezvous with a satellite. How can we be sure that it is possible to achieve the objective, i.e. does there exist at least one control strategy to achieve the objective? This is the *controllability* question.

Suppose next that a system Σ is described by the equations

$$\left.\begin{array}{l} \dot{x} = Ax \\ y = Cx \end{array}\right\} \qquad \begin{array}{l} x(0) = x_0 \\ x \in \mathbb{R}^n, \quad y \in \mathbb{R}^m \end{array}$$

We are given exact measurements of y over some period $[0, T]$. Is it possible to determine x_0 from these measurements? This is the *observability* question.

In later chapters we shall be interested in optimal control. That is, with the question of driving a system to achieve an objective in some best possible sense. Controllability of a system is closely linked with the question of existence of optimal controls. *Optimal state identification* for the system above is concerned with the problem: Given *noisy* measurements of the output y over some period $[0, T]$ and given also that A, C may be subject to uncertainty, form an estimate $\hat{x}_0$ of x_0 such that $\|\hat{x}_0 - x_0\|$ is minimized. We do not deal with state identification in the book because it would require too great an excursion into statistical aspects.

However, observability stands in a similar position to state estimation as does controllability to optimal control. It is in this sense that controllability and observability are dual properties.

The following theorem will be needed to support later results on controllability.

Lemma 7.7. *Let M be an $m \times n$ matrix, $n \leqslant m$, with complex valued entries and define $G = M^*M$. Then G is a positive semi-definite Hermitian matrix.* (The proof of the lemma can be found in texts on matrices.)

Theorem 7.8. *Let F be an $m \times n$ matrix whose elements are measurable functions defined on I, then the row vectors of F constitute a linearly independent set in $R^m(I)$ if and only if the matrix G, defined by*

$$G = \int_I F(t)F^*(t)\,dt \quad \text{is positive definite.}$$

Proof. Let z be a complex valued m vector. By Lemma 7.7, G is positive semi-definite while if $z^*Gz \neq 0$ for any non-zero choice of z then G is positive definite.

$$z^*Gz = \int_I \|z^*F(t)\|^2 \, dt$$

Thus, there exists $z \neq 0$ such that $z^*Gz = 0$ if and only if $z^*F(t) = 0$ for almost all $t \in I$. Linear independence of the rows of $F(t)$ ensures that $z^*F(t) \neq 0$ and G is then positive definite. Conversely, G positive definite implies that $z^*F(t)$ is not identically zero on I and thus the rows of $F(t)$ must constitute a linearly independent set.

Definitions

A system Σ is defined to be *controllable at* t_0 if given t_0, $x(t_0)$ there exists a $t \in I$ and a $u \in \Omega$ such that

$$\phi(x(t_0), t_0, u(\tau), t) = 0$$

If the statement above is true for every $x(t_0) \in X$ the system is said to be *completely controllable at* t_0.

If the system Σ is *completely controllable for every* $t_0 \in I$ then for brevity Σ will be defined simply as *controllable*.

Theorem 7.9. *Let Σ be a controllable system and let $t_0 \in I$, $x_0, x_1 \in X$ be given, then there exists $t_1 \in I$ and u defined on $[t_0, t_1]$ such that*

$$\phi(x_0, t_0, u, t_1) = x_1(t_1).$$

Proof. Put $z = x_0 - \phi^{-1}(x_1, t_0, u, t_1)$, $z \in X$

and

$$\phi(z, t_0, u, t_1) = x_1 - x_1 = 0.$$

Assertion. *The following two statements are equivalent:*

(i) *The system Σ is controllable.*
(ii) *The mapping $\phi: U \to X$ is surjective for all $t_0 \in I$, all $x_0 \in X$.*

Proof. Follows from the definition of controllability.

The following lemma finds application in the proof of theorems on controllability.

Lemma 7.10.

$$\phi(x_0, t_0, u, t_1) = 0$$

if and only if

$$-\Phi(t_1, t_0)x_0 = \int_{t_0}^{t_1} \Phi(t_1, \tau)B(\tau)u(\tau)\,d\tau.$$

(The lemma is a direct consequence of the system axioms).

7.3.2 Controllability of general (time varying) systems

Define the matrix

$$M = \int_{t_0}^{t_1} \Phi(t_1, \tau)B(\tau)B^*(\tau)\Phi^*(t_1, \tau)\,d\tau \tag{7.3.1}$$

M is a square matrix with *constant* coefficients. * denotes complex conjugate.

Theorem 7.11. *The three following statements are equivalent:*

(i) *The system Σ is controllable.*
(ii) *The matrix M is positive definite.*
(iii) *The rows of the matrix $\Phi(t_1, \tau)B(\tau)$ are linearly independent over $[t_0, t_1]$.*

Proof. (ii) $\Rightarrow$ (i).
Assume M is as in equation (7.3.1). Then multiplying by $-M^{-1}\Phi(t_1, t_0)x_0$,

$$-MM^{-1}\Phi(t_1, t_0)x_0 = -\int_{t_0}^{t_1} \Phi(t_1, \tau)B(\tau)B^*(\tau)\Phi^*(t_1, \tau)\,d\tau M^{-1}\Phi(t_1, t_0)x_0$$

$$-\Phi(t_1, t_0)x_0 = \int_{t_0}^{t_1} \Phi(t_1, \tau)B(\tau)u(\tau)\,d\tau$$

where

$$u(\tau) = -B^*(\tau)\Phi^*(t_1, \tau)M^{-1}\Phi(t_1, t_0)x_0.$$

Referring to Lemma 7.10, this condition can be seen equivalent to controllability.

(i) $\Rightarrow$ (ii)
Assume that M is not positive definite, then by Theorem 7.8 the row vectors of the matrix $\Phi(t_1, \tau)B(\tau)$ are not a linearly independent set and the mapping $\phi: U \to X$ is not surjective and the system Σ is not controllable.

(ii) ⇔ (iii)

From Theorem 7.8 positive definiteness of M is equivalent to linear independence of the row vectors of $\Phi(t_1, \tau)B(\tau)$ (Put $F(t) = \Phi(t_1, \tau)B(\tau)$ in Theorem 7.8, then

$$F^*(t) = B^*(\tau)\Phi^*(t_1, \tau)$$

and the theorem is directly applicable).

7.3.3 Controllability of time invariant systems

Theorem 7.12 *The time invariant system Σ is controllable if and only if the matrix*

$$Q_c \triangleq (B \mid AB \mid A^2B \dots A^{n-1}B)$$

satisfies Rank $Q_c = n$, where $X = \mathbb{R}^n$ and hence A is an $n \times n$ constant matrix.

Proof. Controllability is equivalent to the statement: For an arbitrary element $x_1 \in X$ there exist t_1 and u defined on $[0, t_1]$ such that

$$x_1(t_1) = \int_0^{t_1} \Phi(t_1 - \tau)Bu(\tau)\,d\tau$$

$$\Phi(t_1 - \tau) = e^{A(t_1 - \tau)} = e^{At_1}e^{-A\tau}$$

From the Cayley–Hamilton theorem of linear algebra (see for instance reference B11). There exist real numbers γ_i such that

$$\Phi(t_1 - \tau) = \sum_{k=1}^{n} \gamma_k A^{n-k}$$

Hence, if the system is controllable,

$$x_1(t_1) = \int_0^{t_1} \sum_{k=1}^{n} \gamma_k A^{n-k} Bu(\tau)\,d\tau$$

$$= B\int_0^{t_1} \gamma_1 u(\tau)\,d\tau + AB\int_0^{t_1} \gamma_2 u(\tau)\,d\tau + \dots + A^{n-1}B\int_0^{t_1} \gamma_n u(\tau)$$

$$= Q_c \begin{pmatrix} \int_0^{t_1} \gamma_1 u(\tau)\,d\tau \\ \dots\dots\dots \\ \dots\dots\dots \\ \int_0^{t_1} \gamma_n u(\tau)\,d\tau \end{pmatrix}$$

Hence Q_c must have rank n in order that its range contains any arbitrary element $x_i \in X = \mathbb{R}^n$.

To complete the proof assume that the system Σ is not controllable then (see the proof of Theorem 7.8) there exists an element $z \in X$, $z \neq 0$ such that

$$z^T e^{At} B = 0, \qquad \forall t \in I.$$

By successive differentiation at $t = 0$, (this is permissible since e^{At} is analytic on I),

$$z^T A^k B = 0$$

for all integers $k \geqslant 0$ and thus $z^T Q_c = 0$, which implies that Rank $Q_c < n$.

7.3.4 Observability

A particular state x_0 of a system Σ is defined to be unobservable on $[t_0, t_1]$ if

$$y(t) = \eta(\phi(x_0, t_0, 0, t), t) = 0, \quad \forall t \in [t_0, t_1]$$

(Equivalently $C(t)\Phi(t, t_0)x_0 = 0, \forall t \in [t_0, t_1]$).

A system is said to be *completely observable on* $[t_0, t_1]$ if no state is unobservable on $[t_0, t_1]$.

A system that is completely observable will be defined simply as an *observable system.*

Theorems on observability

The theorems are given without proofs. The proofs follow closely those for the analogous theorems on controllability and they can be found in reference K1.

Theorem 7.13. *Let*

$$P = \int_{t_0}^{t_1} \Phi^*(\tau, t_0) C^*(\tau) C(\tau) \Phi(\tau, t_0) \, d\tau$$

Then the following statements are equivalent:

(i) *The system Σ is observable on $[t_0, t_1]$.*
(ii) *The matrix P is positive definite.*
(iii) *The column vectors of $C(t)\Phi(\tau, t_0)$ constitute a linearly independent set.*

Theorem 7.14. *Let Σ be a time-invariant system for which the state space $X = \mathbb{R}^n$ and define a matrix*

$$Q_0 = \begin{pmatrix} C \\ CA \\ CA^2 \\ \vdots \\ CA^{n-1} \end{pmatrix}$$

Then the system Σ is observable if and only if rank $Q_0 = n$.

7.3.5 Controllability and observability: discussion

The details should not be allowed to obscure the simple structural role of these two properties. Refer to Figure 7.3, which illustrates the decomposition theorem of Kalman. Every linear dynamic system can be decomposed into four parts as shown in the figure.

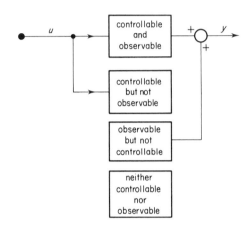

Figure 7.3. Illustrating the decomposition theorem of Kalman.

The properties are important in realization theory. Roughly, only the controllable and observable part of a system can be identified from input–output data. In stability theory, it has already been pointed out, the properties of controllability and observability have an important part to play.

In optimal control, and particularly in time-optimal control, controllability has a geometric interpretation and this is linked to the characterization of optimal controls.

7.4 EXERCISES

(1) Prove that a system is asymptotically stable if and only if

$$\|\Phi(t)\| \to 0 \quad \text{of} \quad t \to \infty$$

where Φ is the transition matrix of the system.

(2) Let $\dot{x} = Ax$ represent a linear time invariant system. Show that if A has a zero eigenvalue then the system may come to equilibrium at points other than the origin.

(3) Let Σ be a time invariant second order system described by the equation $\dot{x} = Ax$. Argue geometrically that if $\langle \dot{x}, x \rangle < 0$ for all x then the system must be stable. Relate this result to the sign properties of the matrix A and then to the eigenvalues of A.

(4) Show by Lyapunov's method that the circuit below is asymptotically stable to the origin in (i, v) space

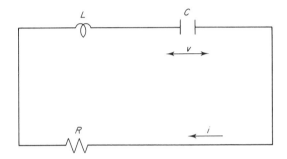

Take current i, voltage v as state variables and take as the Lyapunov function the total energy stored

$$= \tfrac{1}{2}Li^2 + \tfrac{1}{2}Cv^2$$

Assume L, C, R are all positive quantities.

(5) Investigate the stability properties of the system below (Willems)

$$\dot{x} = \begin{pmatrix} 0 & 1 & 0 \\ -1 & 0 & 0 \\ 0 & 0 & -1 \end{pmatrix} x + \begin{pmatrix} 0 \\ 0 \\ 1 \end{pmatrix} u$$

Which of the different stability criteria does the system satisfy?

(6) (Involves further reading.) For single-input single-output systems a number of tests exist to determine whether a system is stable without the need to determine eigenvalues. Three such tests are those of Hurwitz, Routh or Lienard-Chipart.

Read up one of these in a conventional control text and apply it to the problem of question (5).

(7) Let $\dot{x} = Ax$, $x \in X = \mathbb{R}^3$, satisfy $Ax = -\text{grad } \phi$ for some scalar valued function $\phi: X \to \mathbb{R}^1$.

Then the system $\dot{x} = Ax$ is known as a *gradient system*. Show that ϕ is a Lyapunov function.

(8) Let $X = \mathbb{R}^n$ and A be an $n \times n$ matrix with real entries. What are the conditions on A that ensure that $A: X \to X$ is a bijection?

(9) Explain and interpret geometrically the ways in which a system described by the equation $\dot{x} = Ax + Bu$ can fail to be controllable.

Show that the following statements are equivalent for a system described by the equation $\dot{x} = Ax + Bu$, A an $n \times n$ matrix

(a) The system is controllable.

(b) The matrix $\int_{t_1}^{t_2} e^{-A^T \tau} BB^T e^{-A\tau} d\tau$ is nonsingular for any t_1, t_2, with $t_2 > t_1$.

(c) The rows of the matrix $e^{-At}B$ are linearly independent for all $t > 0$.

(d) Rank $(B/AB \ldots | A^{n-1}B) = n$.

(10) Determine whether the system below is (a) controllable, (b) observable:

$$\dot{x} = \begin{pmatrix} -3 & 1 \\ -2 & 1.5 \end{pmatrix} x + \begin{pmatrix} 1 \\ 4 \end{pmatrix} u \qquad y = \begin{pmatrix} 2 & 1 \\ 1 & 4 \end{pmatrix} x$$

(11) A system is described by the equation:

$$\dot{x}(t) = \begin{pmatrix} -2 & 0 \\ 0 & 3 \end{pmatrix} x(t), \qquad x(0) \neq 0$$

$$y(t) = (a \quad b)x(t).$$

The output y of the system is measured at two different times during a transient. Under what conditions on a, b would it be possible to determine the initial state $x(0)$?

(12) A system is described by the equations

$$\dot{x}(t) = \begin{pmatrix} 0 & 6 \\ 1 & -1 \end{pmatrix} x(t) + \begin{pmatrix} b & 1 \\ a & b \end{pmatrix} u(t).$$

The system is known to be uncontrollable; calculate the values of a and b.

CHAPTER 8

Minimum Norm Control

8.1 INTRODUCTION

In control theory, optimization problems are essentially concerned with achieving a given control objective at minimum cost. The cost aspect is made precise by the use of a cost index $J(x, u, t)$ that, often without modification, can be used as a norm on the control space U.

Examples of cost indices that occur naturally in control problems are:

(i) $J = \int_I |u(t)|\, \mathrm{d}t$

(ii) $J = \int_I |u(t)|^2\, \mathrm{d}t$

(iii) $J = \max_{t \in I} (|u(t)|)$

(iv) $J = \int_I |u(t)|^2 + |x(t)|^2\, \mathrm{d}t.$

Cases (i) to (iii) are, respectively, the minimum fuel, minimum energy and minimum peak amplitude criteria while case (iv) is a commonly used criterion in regulator design. Cases (i) to (iii) correspond to the $L^1(I)$, $L^2(I)$, $L^\infty(I)$ norms respectively while case (iv) is a norm for the product space $U x X$. Since in control problems, x depends on u, case (iv) can be brought within the same framework as the first three cases.

An optimal control problem that will recur in this chapter is the following. Determine $u(t)$ such that the system state is brought from a known initial state $x(t)$ to a desired final state z while minimizing a given cost index J.

From a functional analytic viewpoint this optimization problem can then be formulated broadly as: Determine the minimum norm pre-image in $\Omega \subset U$ of a given element z in the state space X. It can be seen that this important class of optimization problems can be formulated naturally as minimum norm problems which can be considered within a function analytic framework.

8.2 MINIMUM NORM PROBLEMS: LITERATURE

An early paper on minimum norm problems was reference (A3) concerned with the so-called L problem. References where problems in control theory have been studied as minimum norm problems include Aoki (A6), Luenberger (L11), Neustadt (N1–N6), Porter (P7–P13), and Swiger (S10). Mizukami (M5) has considered the minimum norm problem for the case where isolated point constraints appear at intermediate times and has given numerical results for a simple example. A practical case where such control might be required occurs in train control where isolated speed limits occur at discrete junctions.

The functional analysis needed in this section is summarized concisely in the book by Day (D7). Specialist material on reflexive spaces can be found in the papers by James (J2, J3). The results on the uniqueness of Hahn–Banach extensions are in a paper by Phelps (P3).

8.3 MINIMUM NORM PROBLEMS: OUTLINE OF THE APPROACH

When the minimum norm problem is formulated in Hilbert space, the problems of existence, uniqueness and characterization of optimal controls are particularly simple. This corresponds to the fact, well known to control engineers, that optimization problems with quadratic cost indices are particularly tractable. The unique decomposition property of Hilbert space and the properties of mappings of finite rank are the main tools required.

The minimum norm problem where the space U is a Banach space is more general and contains the Hilbert space problem as a special case. Existence proofs depend upon the application of the Hahn–Banach theorem in one form or another.

A natural approach to the question of existence of a minimum norm element would be to establish compactness of an appropriate subspace of the space U and then appeal to the Weierstrass theorem. However, since U is an infinite dimensional space, closed balls in U fail to be compact. For this reason, the weak and weak* topologies for U are introduced to allow the concept of weak compactness to be used. Reflexivity and the Bourbaki–Alaoglu theorem are required in the development.

Uniqueness of the minimum norm control requires the concepts of rotundity and smoothness and it is also possible to appeal to theorems of Phelps on the uniqueness of Hahn–Banach extensions.

A characterization of the minimum norm control in U, where U is a Banach space, can be derived from the condition for equality in the Hölder inequality.

8.4 MINIMUM NORM PROBLEM IN HILBERT SPACE: DEFINITION

8.4.1 Problem definition

Let Σ be a controllable system with finite dimensional state space. Let $t_0 \in I$, $x(t_0) \in X$, $z \in X$ be given. Let U be the Hilbert space $L^2(I)$.

The minimum norm control problem is then to determine $u(t) \in U$ such that for some $t_1 \in I$;

(i) $x(t_1) = z$;
(ii) $\|u\|$ is minimized on $[t_0, t_1]$ where $\|\cdot\|$ represents the norm on $L^2[t_0, t_1]$.

8.4.2 Existence and uniqueness of a minimum norm element

Theorem 8.1. *A unique minimum norm control exists for the problem of Section 8.4.1.*

Proof. There exists $u(t)$ defined on $[t_0, t_1]$ such that $x(t_1) = z$; this follows from the controllability assumption. With t_0, t_1 fixed, $x(t_0)$ fixed, represent the mapping

$$\phi: X \times I \times U \times I \to X$$

by the mapping S,

$$S: U \times I \to X$$

The set $\{S^{-1} z\} \subset U$ is therefore non-empty.

Let $\{e_i\}$ be a basis for X then z can be expressed

$$z = \{\alpha_1 e_1 + \ldots + \alpha_n e_n\}$$

where the α_i are numbers. If $Su = z$ then Su can be expressed

$$Su = \sum_{i=1}^{n} f_i(u)$$

where the f_i are linearly independent functionals on U. The f_i generate an n dimensional subspace, say $L \subset U$. Since U is a Hilbert space it can be decomposed as $U = L \oplus L^\perp$. Since each $f_i \in L$ we must have $f_i(u) = 0$ for every $u \in L^\perp$. Hence $L^\perp \subset N_S$ (where N_S is the nullspace of S).

Assume now that $u \in N_S$, then $Su = 0$ and this implies that $f_i = 0, i = 1, \ldots, n$. Thus

$$N_S \subset L^\perp$$

from which

$$N_S = L^\perp$$

Since L, X both have dimension n, the mapping S restricted to L denoted by S_L is injective and a unique minimum norm element $S_L^{-1}z, L \subset U$, exists.

(Let $u = u_1 + u_2$ be any u for which $Su = z$ with $u_1 \in L$, $u_2 \in L^\perp$, $u_2 \neq 0$, then $Su = Su_1 + Su_2 = Su_1$ while $\|u\| = \|u_1\| + \|u_2\|$, so that $\|u\| > \|u_1\|$ and $u_1 \in L$ is indeed the minimum norm element satisfying $Su = z$).

Comment

In the above proof, let $u(t)$ be any solution of the equation $Su = z$ then the unique minimum norm control is the projection of $u(t)$ onto the closed subspace $L(f_1, \ldots, f_n)$.

The n linear functionals f_i can be obtained from the system matrices associated with the system Σ. Let $\Psi(t)$ be the system matrix for Σ defined as in Section 6.3, then each row of the matrix corresponds to one of the required n linear functionals f_i. The ideas just expressed can be used as a basis for the characterization of minimum norm elements in a Hilbert space. This characterization is considered in the next sections.

8.4.3. Characterization of the minimum norm control in L^2

Let S be a linear transformation $S : U \to X$ where U is a Hilbert space and X is $\mathbb{R}^n$. Given $z \in X$ determine the $u_z \in U$ that is the minimum norm element satisfying $Su_z = z$. S can be represented by a square matrix that can be written

$$S = \sum_{i=1}^{n} e_i d_i^T$$

where $\{e_i\}$ is an orthonormal basis for X and $\{d_i\}$ is a basis for the n dimensional subspace L of U to which the unique element u_z is known to belong, from results in the previous section. Each d_i must be a scalar valued function on U, i.e. $\{d_i\}$ is a basis of functionals. z and u_z can each be represented in terms of basis vectors:

$$z = \sum_{i=1}^{n} \beta_i e_i, \qquad u_z = \sum_{i=1}^{n} \alpha_i d_i, \qquad \beta_i, \alpha_i \text{ real numbers.}$$

The relation $z = Su_z$ can now be written

$$\sum_{i=1}^{n} \beta_i e_i = \left(\sum_{i=1}^{n} e_i d_i \right)^T \left(\sum_{i=1}^{n} \alpha_i d_i \right)$$

Then

$$\beta_q = \sum_{i=1}^{n} \alpha_i \langle d_q, d_i \rangle$$

Leading to

$$\begin{pmatrix} \beta_1 \\ \vdots \\ \vdots \\ \beta_n \end{pmatrix} = \left(D_{ij} = \langle d_i, d_j \rangle \right) \begin{pmatrix} \alpha_1 \\ \vdots \\ \vdots \\ \alpha_n \end{pmatrix}$$

or

$$\beta = D\alpha \qquad (8.4.1)$$

The matrix transforms the coordinates of u_z into the coordinates of z:

$$D : L \rightarrow X$$

Equation (8.4.1) offers a method for the coordinates $\alpha_i(t)$ of the desired control function to be determined.

In solving a concrete optimization problem the required control can easily be found once the basis $\{d_i\}$ is known. The route to determining the $\{d_i\}$ is through the system transition matrix as outlined below. Assume that a system described by the vector-matrix equations

$$\dot{x} = Ax + Bu \qquad (8.4.2)$$

is to be driven from an initial state $x(t_0)$ to a desired state z at time t_f while minimizing a cost index J of the form

$$J = \int_{t_0}^{t_f} |u(\tau)|^2 \, dt$$

The solution of (8.4.2) is

$$x(t_f) = \Phi(t_f, t_0) x(t_0) + \int_{t_0}^{t_f} \Phi(t_f, \tau) Bu(\tau) \, dt$$

so that the control problem is to ensure that the following equation is satisfied:

$$z' \triangleq z - \Phi(t_f, t_0)x(t_0) = \int_{t_0}^{t_f} \Phi(t_f, \tau)Bu(\tau)\, d\tau = Su$$

with u_z being the u that satisfies this equation and minimizes the given cost function.

Each element $(Su)_i$ can be written

$$(Su)_i = \sum_{k=1}^{n} \int_{t_0}^{t_f} (\Phi(t_f, \tau)B)_{ik} u_k(\tau)\, d\tau$$

Let $\{e_i\}$ be the usual coordinate basis for X then

$$Su = \sum_{i=1}^{n} e_i(Su)_i$$

$$= \sum_{i=1}^{n} e_i \sum_{k=1}^{n} \int_{t_0}^{t_f} (\Phi(t_f, \tau)B)_{ik} u_k(\tau)\, d\tau$$

The rows of $(\Phi(t_f, \tau)B)$, $i = 1, \ldots, n$ are the functionals that form a basis for the subspace L of U.

The matrix D in equation (8.4.1) is interpreted by

$$D = \int_{t_0}^{t_f} (\Phi(t, \tau)B)(\Phi(t, \tau)B)^*\, d\tau \qquad (8.4.3)$$

Let

$$\beta = \begin{pmatrix} \beta_1 \\ \cdot \\ \dot{\beta}_n \end{pmatrix}$$

be the coordinates of z' then in the orthonormal Euclidean basis for X,

$$\beta = z' = z - \Phi(t_f, t_0)x(t_0) \qquad (8.4.4)$$

$$u_z = \sum_{i=1}^{n} \alpha_i d_i = \sum_{i=1}^{n} \alpha_i \sum_{k=1}^{n} (\Phi(t_f, \tau)B)_{ik}$$

$$= (\Phi(t_f, \tau)B)^*\alpha \qquad (8.4.5)$$

Using (8.4.5), (8.4.4), (8.4.3) and (8.4.1)

$$u_z(t) = (\Phi(t_f, \tau)B)^* D^{-1}\beta$$

$$= (\Phi(t_f, \tau)B)^* D^{-1}(z - \Phi(t_f, t_0)x(t_0))$$

$$u_z(t) = (\Phi(t_f, t)B)^*\left(\int_{t_0}^{t_f} (\Phi(t_f, \tau)B)(\Phi(t_f, \tau)B)^*\, d\tau \right)^{-1} (z - \Phi(t_f, t_0)x(t_0)) \qquad (8.4.6)$$

It is clear that, given a mathematical model in the form of equation (8.4.2) the complete set of vector time functions $u_z(t)$ can be calculated for the period $[t_0, t_f]$.

Alternatively, it can also be seen that t_0 could be regarded always as the present time, that $x(t_0)$ could be measured and the optimal control calculated as a function of $(t_f - t_0)$, the time remaining. This second approach has an element of feedback correction that allows the use of an approximate model.

8.4.4 Characterization of the minimum norm control in Hilbert space: simple numerical example

Let Σ be a system that can be represented by the equations

$$\begin{pmatrix} \dot{x}_1 \\ \dot{x}_2 \end{pmatrix} = \begin{pmatrix} 0 & 1 \\ 0 & -1 \end{pmatrix} \begin{pmatrix} x_1 \\ x_2 \end{pmatrix} + \begin{pmatrix} 0 \\ 1 \end{pmatrix} u$$

where $I = \mathbb{R}^1$ and

$$x(t) \in \mathbb{R}^2 = X, \qquad u(t) \in L^2(\zeta), \quad \zeta \subset I$$

A control problem is defined: Given $z(T) \in X$ determine $u(t)$ in $L^2(\zeta)$ such that

(i) $x(T) = z(T)$
(ii) $\|u\|$ is minimized

Let

$$x(0) = \begin{pmatrix} 0 \\ 0 \end{pmatrix}, \qquad z = \begin{pmatrix} \gamma \\ 0 \end{pmatrix}, \qquad T = 1, \qquad \zeta = [0, 1]$$

$$\Phi(t) = \begin{pmatrix} t & (1 - e^{-t}) \\ 0 & e^{-t} \end{pmatrix} \tag{8.4.7}$$

$u_{opt}(t)$ is determined from (8.4.6). Noting that for a time invariant system $\Phi(t_f, \tau)$ becomes a function of the single argument $t_f - \tau$. Putting $t = t_f - \tau$ $t_f = 1$ in (8.4.7) yields

$$\Phi(t_f, \tau)B = \begin{pmatrix} 1 - e^{-(1 - \tau)} \\ e^{-(1 - \tau)} \end{pmatrix}$$

Then

$$u_{opt}(t) = (1 - e^{t-1}, e^{t-1}) \left[\int_0^1 \begin{pmatrix} 1 - e^{\tau-1} \\ e^{\tau-1} \end{pmatrix} (1 - e^{\tau-1}, e^{\tau-1}) \, d\tau \right]^{-1}$$

$$\left[\begin{pmatrix} \gamma \\ 0 \end{pmatrix} - \begin{pmatrix} 1 & 1 - e^{-1} \\ 0 & e^{-1} \end{pmatrix} \begin{pmatrix} 0 \\ 0 \end{pmatrix} \right]$$

and after some manipulation

$$u_{opt}(t) = \gamma(1 + e - 2e^t)/(3 - e)$$

Before proceeding to study minimum norm control problems in Banach spaces, a number of results will be established.

8.5 MINIMUM NORM PROBLEMS IN BANACH SPACE

8.5.1 Preliminaries

Theorem 8.2. A consequence of the Hahn Banach theorem. *Let X be a normed linear space and let $x \in X$, $x \neq 0$, then there exists $f \in X^*$ such that $f(x) = \|x\|$ and $\|f\| = 1$.*

Proof. [We first note that if X is an inner product space, the theorem is true. For let

$$f(x) = \langle x, y \rangle \qquad \text{where } y = x_1/\|x_1\|$$

Then

$$f(x_1) = \|x_1\| \quad \text{and} \quad \|f\| = 1.]$$

Given $x \in X$, denote by $L(x)$ the subspace generated by x. Define a functional on this one dimensional subspace by

$$f(\alpha x) = \alpha \|x\|$$

Every element in $L(x)$ is of the form αx. Now

$$f(x) = \|x\| \quad \text{and} \quad \|f\| = \sup \frac{f(x)}{\|x\|} = 1$$

By the Hahn–Banach theorem, f can be extended to a functional defined on the whole of X and satisfying $\|f_x\| = \|f\|$ on $L(x)$ (where f_x is the extended functional).

Extrema: definitions and theorems

Let X be a Banach space and let X^* be the dual of X. Let β be the unit ball in X and β^* be the unit ball in X^*.

Given $x \in X$; $x \neq 0$: If there exists $\tilde{f} \in \beta^*$ such that $f(x) = \|x\|$, $\|\tilde{f}\| = 1$, then $\tilde{f}$ is called an *extremum* for x.

Given $f \in X^*$; $f \neq 0$: If there exists $\tilde{x} \in \beta$ such that $f(\tilde{x}) = \|f\|$, $\|\tilde{x}\| = 1$, then $\tilde{x}$ is called an extremum for f.

Existence and uniqueness of extrema

Let X be a normed linear space then by Theorem 8.2, given $x \in X$ there exists at least one extremum for x. Similar reasoning can be applied to X^* so that given $f \in X^*$ there exists at least one extremum $x^{**} \in \beta^{**}$ for f. This functional can be written $x^{**}(f)$ or $\langle f, x^{**} \rangle$. Now assume that X is reflexive, then there exists x such that

$$\langle x, f \rangle = \langle f, x^{**} \rangle$$

for every $f \in X^*$.

Theorem 8.3

(i) *If X is reflexive then given $f \in X^*$ there exists at least one extremum for f in $\beta \in X$.* (Proved above).

(ii) *If X is rotund then to each element $f \in X^*$, $f \neq 0$, there corresponds at most one extremum in $\beta \subset X$ for f.*

(iii) *If X is smooth then each $x \in X$ has at most one extremum in $\beta^* \in X^*$.*

Proof of (ii). Suppose $x_1, x_2 \in \beta$ such that $f(x_1) = f(x_2) = \|f\|$ and

$$\|x_1\| = \|x_2\| = 1$$

Put $x = (x_1 + x_2)/2$. Then from these last relations

$$f(x) = \|f\| \quad \text{and} \quad \|x\| \leqslant 1.$$

But $f(x) \leqslant \|f\| \|x\|$, from which $\|x\| \geqslant 1$, so that $x \in \partial\beta$. But x is the midpoint of the segment $[x_1; x_2]$ and since X is rotund, cannot be in the boundary of β.

Proof of (iii). By the definition of smoothness, through each point $x_1 \in \partial\beta$ there passes a unique support hyperplane H, supporting β at x_1. Every hyperplane is the translation of the null space of a linear functional f so that H can be written

$$H = \{x \mid f(x) = \lambda\} \qquad \text{a real number.}$$

If the hyperplane is to contain x_1 then we must have $f(x_1) = \|x_1\|$ and thus $f(x) = 1$ generates the hyperplane H.

f is unique, for otherwise, suppose g is a second functional satisfying the requirement

$$H = \{x \mid g(x) = 1\}$$

This would lead to the condition

$$1 = \langle x, g \rangle = \langle x, f \rangle, \qquad \forall x \in X$$

$$\langle x, g \rangle - \langle x, f \rangle = 0$$

$$\langle x, g - f \rangle = 0, \qquad \forall x \in X$$

$$g - f = 0$$

Summary—existence and uniqueness of extrema

	Existence condition	*Uniqueness condition*
$x \in X$	X any normed linear space	X rotund
$f \in X^*$	X reflexive	X smooth

8.5.2 Definition of problem

The case where the state space $X = \mathbb{R}^1$ is considered. There is no difficulty in extending the results to the case where $X = \mathbb{R}^n$.

Let a minimum norm problem be posed as follows: Let Σ be a single state linear dynamic system. Let U be a Banach space, with t_0, t_1 fixed, $x(t_0)$ fixed; let $S : U \to X$ represent the mapping ϕ. Since $X = \mathbb{R}^1$, Su can be represented by a linear continuous functional $\langle v, u \rangle$, where $u \in U$ and where $v \in V$. $V = U^*$ or $U = V^*$.

The minimum norm problem is then, given $z \in X$ determine u on $[t_0, t_1]$ such that:

(i) $x(t_1) = z$.
(ii) $\|u\|$ is minimized, where $\|\cdot\|$ represents the norm on $L^p[t_0, t_1]$.

8.5.3 Existence

Theorem 8.4. *Let $v \in V$ where V is a real normed linear space and let $u \in U$ where $U = V^*$, then there exists a solution to the minimum norm problem.*

Proof. From Section 8.5.1, or by the Hahn–Banach theorem, given $v \in V$ there exists $\tilde{u} \in V^*$ such that

$$\tilde{u}(v) = \|v\|, \qquad \tilde{u} \in \beta^* \subset V^*$$

To be a minimum norm control, u must satisfy the following inequality with equality:

$$|z| = |u(v)| \leqslant \|u\| \, \|v\|$$

Put

$$u(v) = \frac{|z|}{\|v\|}\, \tilde{u}(v)$$

where $\tilde{u}(v)$ is the extremal in V^* for v, then the right side of the inequality becomes

$$\frac{|z|}{\|v\|}\, \|\tilde{u}\|\, \|v\| = |z|$$

since $\|\tilde{u}\| = 1$ and the inequality is satisfied with equality.

Theorem 8.5. *Let* $v \in V$ *where* $V = U^*$ *and let* U *be a reflexive Banach space, then there exists a solution to the minimum norm problem.*

Proof. From Section 8.5.1, given $v \in U^*$ there exists $\tilde{u} \in U$ such that

$$\tilde{u}(v) = \|v\|, \qquad \tilde{u} \in \beta \subset U$$

The remainder of the proof is identical to that of Theorem 8.4.

If the control space U is reflexive then existence of a minimum norm control is guaranteed. However, reflexivity is a very strong requirement and Theorem 8.4 shows that provided U is the normed dual of a real normed linear space, the existence of a minimum norm control can still be guaranteed. This is interesting since it allows for possible manipulation of the original problem so that U becomes a dual space even though in the original formulation it was perhaps the primal space.

Pursuing this argument, consider the "cost indices" that are most commonly used in control theory to ascribe a cost to the control function $u(t)$. These are as follows:

$$J_1 = \int_0^T |u(t)|\, dt$$

$$J_2 = \int_0^T |u(t)|^2\, dt$$

$$J_3 = \sup_{t \in [0,\, T]} \{|u(t)|\}$$

J_2 makes the control space U into a Hilbert space so that existence of a minimum norm control is obviously guaranteed.

J_1 leads to the control space $U = L^1[0, T]$.
J_3 leads to the control space $U = L^\infty[0, T]$.

In neither of these important cases (J_1, J_3) is the control space U reflexive and it appears that minimum norm controls may not exist.

In the case where the norm on U arises from the cost function J_3, the primal space V can be taken as $L_1[0, T]$ so that $V^* = L^\infty[0, T] = U$. Where the norm on U arises from the cost function J_1, U would not then be the dual of any normed space. However, U can perhaps be taken as the set of functions of bounded variation on $[0, T] = BV[0, T]$ so that $V = C[0, T]$ (the set of functions continuous on $[0, T]$) and $U = V^*$.

Before making the manipulations implied above it will be necessary to check that the assumptions on the element $\Psi \in V$ are valid. In particular, the assumption that $V = C[0, T]$ may not be valid in some cases.

From Theorems 8.4 and 8.5 the main existence theorem follows immediately.

Theorem 8.6. *A solution exists to the minimum norm problem in Banach space if either:*

(i) *The space U is reflexive, or*
(ii) *The space U is the normed dual of a real normed linear space.*

In case (i) the minimum norm problem is formulated so that the control space U is the primal space, while in case (ii) U is a dual space.

8.5.4 Uniqueness

Theorem 8.7. *Let $v \in V$, where V is a smooth space, and let $u \in U$, where $U = V^*$, then there exists at most one solution to the minimum norm problem.*

Proof. From Section 8.5.1 to each given $v \in V$ there corresponds at most one extremum,

$$\tilde{u} \in \beta^* \subset V^* = U$$

and thus there is at most one element,

$$\frac{|z|}{\|v\|} \tilde{u}(v)$$

that is a minimum norm control.

Theorem 8.8. *Let $u \in U$ where U is a rotund space and let $V = U^*$ then there exists at most one solution to the minimum norm problem.*

Proof. From Section 8.5.1, given $v \in V$ there exists at most one extremum

$\tilde{u} \in \beta \subset U$ for v. Thus, there is at most one element giving minimum norm control. Combining the results yields the main uniqueness theorem.

Theorem 8.9. *If a minimum norm solution exists for the problem given in Section 8.5.2, this solution will be unique:*
In case (i) *if the space U is rotund.*
In case (ii) *if the space U is the normed dual of a smooth space.*

Comment

Results on reflexivity, rotundity and smoothness are collected together in Chapter 5. Rotundity and smoothness are closely related and from Section 5.6.

$$\text{Rotundity of } U^* \Rightarrow \text{smoothness of } U$$

while if the space U if reflexive, rotundity and smoothness become dual properties such that

$$U^* \text{ rotund} \Leftrightarrow U \text{ smooth}$$

$$U^* \text{ smooth} \Leftrightarrow U \text{ rotund.}$$

8.5.5 Minimum norm problems in Banach spaces: discussion

Clearly minimum norm problems in Banach spaces are considerably more complex than their counterparts in Hilbert space.

The properties required of the space U to ensure existence and uniqueness of a minimum norm control can be summarized.

Theorem 8.10. *A minimum norm problem in Banach space has a unique solution if*

Case (i) (where U is the primal space). *The space U is reflexive and rotund.*
Case (ii) (where U is the dual space). *The space U is the dual of a smooth linear space V.*

It will be useful to determine whether category (ii) actually increases the choice of control spaces for which the minimum norm problem possesses a unique solution.

From Section 8.5.4, (i) $\Rightarrow$ (ii), while, if V is reflexive, (ii) $\Rightarrow$ (i). Thus, (ii) differs from (i) only in the event that V is a smooth but non-reflexive space.

The uniqueness theorems can be confirmed by reference to papers on the uniqueness of Hahn–Banach extensions. For example, Phelphs (P3) gives the following result attributable to Taylor and Foguel.

Theorem 8.11. *Let E be a normed linear space and let M be any subspace of E, then each linear functional on M has a unique norm preserving extension to E if and only if E* is rotund.*

Proof. See reference P3.

Phelps goes on to show that even though the space L^1 does not possess the above uniqueness property, certain subspaces of L^1 do have a unique extension. In contrast, no subspace in L^∞ can be extended uniquely to the whole space.

8.5.6 Minimum norm problem in a Banach space when the state space X is n dimensional

Let Σ be a controllable system. Let U be a Banach space. Let X be $\mathbb{R}^n$. Let $S : U \to X$ be the mapping ϕ with t_0, t_1, $x(t_0)$ fixed. Let $z \in X$ be given. The minimum norm problem is to determine u on $[t_0, t_1]$ such that
   (i) $x(t_0) = z$,
   (ii) $\|u\|$ is minimized, where $\|\cdot\|$ is the norm on $L^p[t_0, t_1]$.

Theorem 8.12. *Let U be a reflexive rotund space; then there exists a unique solution to the minimum norm problem.*

Proof. S is linear and hence $M = \{S^{-1}z\}$ is convex in the rotund space U, by Section 5.5, M contains at most one element of minimum norm.
   M is non-empty since Σ is controllable. M is closed by the continuity of S. Let

$$\alpha = \inf\{\|u\| \,|\, u \in M\}$$

Let

$$K_n = \{u \,|\, u \in M, \|u\| \leqslant \alpha + (1/n)\}, \qquad n \in Z^+$$

$\{K_n\}$ is a decreasing sequence of closed bounded sets—each K_n is non-empty, hence (see Section 5.4) the sequence has non-empty intersection and there exists at least one element u of minimum norm.

8.5.7. Formation of a norm for the space U when there are r control inputs

Let $U = L^p(\tau) \times L^p(\tau) \times \ldots \oplus L^p(\tau)$; r times $\triangleq (L^p)^r (\tau)$, then a product norm can be defined in U by

$$\|u\| = \left[\int_0^T \sum_{i=1}^r |u_i(\tau)|^p \, dt \right]^{1/p}, \qquad p \in [1, \infty)$$

Such a norm is meaningful in the formulation of optimization problems since it can be related to the total energy or cost associated with the input u.

8.6 MORE GENERAL OPTIMIZATION PROBLEMS

Problems with cost index of the form $J = J(x, u)$, for instance the cost index (iv) in Section 8.1, cannot be formulated directly as minimum norm problems on the space U. Further manipulation is required as shown below.

Let Σ be a single state linear dynamic system. Let U be a Banach space $L^p(I)$. Let $X = \mathbb{R}^1$. With t_0, $t_1 \in I$ fixed, $x(t_0)$ fixed. Let $S: U \to X$ represent the mapping ϕ.

Since $X = \mathbb{R}^1$, Su can be represented by a linear continuous functional $\langle v, u \rangle$ where

$$u \in U \quad \text{and} \quad v \in V; \quad V = U^* \quad \text{or} \quad U = V^*$$

Define by T the mapping ϕ with t_0, $x(t_0)$ fixed.

$$T: U \times I \to X$$

Then for all $t \in I$, $x(t) = Tu(t)$.

A minimum norm problem can be formulated: Determine u on $[t_0, t_1]$ such that:

(i) $x(t_1) = z$.

(ii) $\|(u, Tu)\|$ is minimized.

$\{u(t), Tu(t) | t \in I\}$ is the *graph* of $T = G_T$. G_T is a Banach space that can be normed in accordance with the cost index $J(x, u)$. Finally, a mapping Q is defined such that

$$Q(u, Tu) = Su, \quad Q: G_T \to X$$

Lemma 8.13. *Q is a continuous linear functional on G_T.*

Proof.

$$Q(u, Tu) = Su = x \in X$$

$$\|Q(u, Tu)\| = \|Su\| \leqslant \|S\| \|u\| \leqslant \|S\| \|(u, Tu)\|$$

$$\frac{\|Su\|}{\|(u, Tu)\|} \leqslant \|S\|, \quad \|Q\| \leqslant \frac{\|Su\|}{\|(u, Tu)\|} \leqslant \|S\|$$

Q is bounded since S is bounded. Q is linear since

$$Q(u_1 + u_2, T(u_1 + u_2)) = S(u_1 + u_2) = Su_1 + Su_2$$

$$= Q(u_1, Tu_1) + Q(u_2, Tu_2) \qquad \forall u_1, u_2 \in U$$

$$Q(\alpha u, T\alpha u) = S\alpha u = \alpha Su = \alpha Q(u, Tu)$$

Theorem 8.14. *The optimization problem has a unique solution if*

(i) *The subspace* $G_T \in U \times X$ *is reflexive and rotund*; or
(ii) G_T *is the dual of a smooth linear space V.*

Proof. Q is a continuous linear functional and the conditions that z shall have a unique minimum norm pre-image under Q are precisely the same as those for Theorem 8.12.

Theorem 8.15. *Let* U, X *be reflexive rotund Banach spaces then* $U \times X$ *is a reflexive rotund Banach space for which a norm can be defined.*

$$\|(u, x)\| = [\|u\|^p + \|x\|^p]^{1/p}, \qquad 1 < p < \infty$$

Proof. A finite product of reflexive spaces is reflexive and a norm $\|\|\cdot\|\|$ defined

$$\|\|x\|\| = \left(\sum_{i=1}^{n} \|x_i\|^p \right)^{1/p}, \qquad 1 < p < \infty$$

preserves rotundity provided that the norm $\|\cdot\|$ is rotund on each of the contributing spaces.

Theorem 8.16. *Let* U, X *be reflexive rotund Banach spaces and let* $T : U \to X$ *be a bounded bijective mapping then* G_T *is a closed subspace of* $U \times X$, *hence* G_T *is a rotund reflexive Banach space.*

Proof. We wish to show that the limit of a sequence (u_n, T_n) belongs to G_T. Let

$$\lim_{n \to \infty} (u_n, Tu_n) = (u, x)$$

By the continuity of T

$$u_n \to u, \qquad Tu_n \to Tu$$

hence $x = Tu$ and the limit of the sequence is a member of G_T by definition, so that G_T is closed.

Examples of norms on G_T arising from particular cost indices

(i) A norm of the form

$$(\|u\|^p + \|Tu\|^p)^{1/p} = \left(\int_I |u(t)|^p + |x(t)|^p \, dt \right)^{1/p}$$

can be expressed as a norm on the graph of $T = G_T$,

$$(\|u\|^p + \|Tu\|^p)^{1/p} = \|(u, Tu)\| = \|y\|$$

where $y \in G_T$

(ii) Consider a norm of the form

$$J = (\|u - \hat{u}\|^p + \|Tu - \hat{x}\|^p)^{1/p}$$

Let $\hat{w} = (\hat{u}, \hat{x})$, $w = (u, Tu)$. Then the norm can be expressed

$$J = \|w - \hat{w}\|$$

or if we define $y = \hat{w}$, $y \in G_T$, then

$$J = \|y\|$$

where we have assumed that $\hat{w} \in G_T$.

Comment

Once a norm has been defined on the graph of $T = G_T$, the more general optimization problem can be treated by methods exactly parallel to those described for the problems where the cost index was simply a functional defined on the space U.

8.7 MINIMUM NORM CONTROL: CHARACTERIZATION, A SIMPLE EXAMPLE

Let Σ be a single input–single output linear dynamic system, i.e.

$$U = \mathbb{R}^1, \qquad X = \mathbb{R}^1, \qquad I = \mathbb{R}^1.$$

Then an optimal control problem can be presented: Given x_0, t_0, t_1, z, determine $u(t) \in U$ such that

(i) $x(t_1) = z$.
(ii) $\|u\|_p$ is minimized $(1 < p < \infty)$.

Now

$$x(t) = \Phi(t)x_0 + \int_{t_0}^{t_1} \Phi(t_1, \tau)B(\tau)u(\tau)\,d\tau$$

If condition (i) is to be satisfied then we must have

$$e = z \ - \ \Phi(t_1)x_0 = \int_{t_0}^{t_1} \Phi(t_1, \tau)B(\tau)u(\tau)\,d\tau \tag{8.7.1}$$

where e is defined by equation (8.7.1). Define

$$\Psi(t) = \Phi(t_1, t)B(t), \qquad t_1 \in I \text{ fixed.}$$

Then

$$e = \int_{t_0}^{t_1} \Psi(\tau)u(\tau)\,d\tau \qquad (8.7.2)$$

Equation (8.7.1) must be satisfied if condition (i) is to be satisfied.
   Turning to condition (ii), from

$$|e| = \left| \int_{t_0}^{t_1} \Psi(\tau)u(\tau)\,d\tau \right| \leqslant \int_{t_0}^{t_1} |\Psi(\tau)u(\tau)|\,d\tau \qquad (8.7.3)$$

and from the Hölder inequality applied to (8.7.3),

$$\int_{t_0}^{t_1} |\Psi(\tau)u(\tau)|\,d\tau \leqslant \|\Psi\|\,\|u\| \qquad (8.7.4)$$

where it has been assumed that $\Psi(t) \in L^q(I)$.
   From (8.7.3), (8.7.4) the minimum norm control, if it exists, will satisfy

$$\|u\| = \frac{|e|}{\|\Psi\|} \qquad (8.7.5)$$

For (8.7.5) to be satisfied, the inequalities must be satisfied with equality.
This leads to the conditions, from (8.7.3),

$$\text{sign}(u(t)) = \text{sign}(\Psi(t)), \qquad \forall t \in [t_0, t_1] \qquad (8.7.6)$$

and from (8.7.4),

$$|\Psi(t)|^q = \alpha|u(t)|^p, \qquad \forall t \in [t_0, t_1], \quad \alpha \text{ an arbitrary constant.} \qquad (8.7.7)$$

From (8.7.6), (8.7.7),

$$|u(t)| = \alpha^{-1/p}|\Psi(t)|^{q/p} = k|\Psi(t)|^{q/p}$$

$$u(t)\,\text{sign}(u(t)) = k|\Psi(t)|^{q/p}$$

$$u(t) = k\,\text{sign}(\Psi(t)) \cdot |\Psi(t)|^{q/p} \qquad (8.7.8)$$

By substituting equation (8.7.8) into equation (8.7.2), the constant k can be
determined.
   In the special case when $p = q = 2$, equation (8.7.3) becomes

$$e = \langle \Psi, u \rangle$$

where $\langle\ ,\ \rangle$ denotes the inner product in L^2.

From (8.7.8), $u(t) = k\Psi(t)$. From (8.7.2), (8.7.8),

$$e = \int_{t_0}^{t_1} k\Psi(\tau)\Psi(\tau)\,d\tau = k\|\Psi\|^2$$

and thus $k = e/\|\Psi\|^2$ yielding the optimal control law

$$u(t) = e\Psi(t)/\|\Psi\|^2$$

8.8 DEVELOPMENT OF NUMERICAL METHODS FOR THE CALCULATION OF MINIMUM NORM CONTROLS

Let Σ be a dynamic system with $U = \mathbb{R}^r$, $X = \mathbb{R}^n$, $I = \mathbb{R}^1$. Given x_0, t_0, t_1, z determine $u(t) \in U$ such that

(i) $x(t_1) = z$.
(ii) $\|u\|_p$ is minimized, $1 \leqslant p \leqslant \infty$.

This problem is an obvious generalization of that of Section 8.7. By a development parallel to that in Section 8.7 we obtain the relation (cf. equations (8.7.1), (8.7.2)).

$$e = z - \Phi(t_1)x_0 = \int_{t_0}^{t_1} \Psi(t_1 - \tau)u(\tau)\,d\tau \qquad (8.8.1)$$

$u(t)$ must be chosen on the interval $[t_0, t_1]$ to make equation (8.8.1) true; e is a vector. Consider its ith component e_i:

$$e_i = \int_{t_0}^{t_1} \psi_i(t_1 - \tau)u(\tau)\,d\tau = \langle \psi_i, u \rangle = f(\psi_i)$$

where ψ_i is the ith row of the matrix Ψ and f is the unique functional corresponding to the inner product.

Let λ_i be an arbitrary scalar then

$$\lambda_i e_i = \lambda_i f(\psi_i) = f(\lambda_i \psi_i)$$

$$\sum_1^n \lambda_i e_i = \sum_1^n f(\lambda_i \psi_i) = \langle \lambda, e \rangle \qquad (8.8.2)$$

λ being a vector with arbitrary elements $\lambda_1, \ldots, \lambda_n$. If equation (8.8.2) is true for at least n different linearly independent then equation (8.8.1) will also be true.

By Hölders inequality,

$$\left| f\left(\sum_1^n \lambda_i \psi_i \right) \right| \leqslant \|u\|_p \left\| \sum_1^n \lambda_i \psi_i \right\|_q \qquad (8.8.3)$$

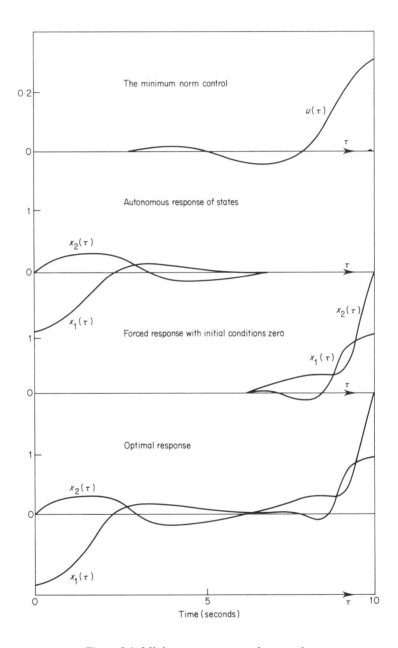

Figure 8.1. Minimum norm control: example.

From equations (8.8.2), (8.8.3),

$$\|u\|_p \geq \frac{|\langle \lambda, e \rangle|}{\left\|\sum_{i}^{n} \lambda_i \psi_i\right\|_q} \tag{8.8.4}$$

Every control driving the system to the point z must satisfy equation (8.8.4) while for optimal (minimum norm) control it can be shown that equation (8.8.4) must be satisfied with equality (see reference (K13)).

Numerically, we must search for an n vector λ such that the right side of equation (8.8.4) takes on its maximum. Let λ^* be the vector that maximizes the right side of equation (8.8.4) and define

$$e = \sum_{1}^{n} \lambda_i^* \psi_i(t_i - \tau)$$

By methods of ordinary calculus it is not difficult to show that each element of the required norm control is given by

$$u_j(\tau) = \frac{\|e(\tau)\|^{q-1} \operatorname{sign}(e(\tau))}{(\|e\|_q)^q} \tag{8.8.5}$$

Numerical example

A system is described by the equation

$$\dot{x} = \begin{pmatrix} 0 & 1 \\ -1 & -1 \end{pmatrix} x + \begin{pmatrix} 0 \\ 1 \end{pmatrix} u$$

$$x(0) = \begin{pmatrix} -0.9 \\ 0 \end{pmatrix}, \qquad z = \begin{pmatrix} 1 \\ 2 \end{pmatrix}, \qquad t_1 = 10 \text{ seconds, } p = 2.$$

The minimum norm control is to be found to drive the system from $x(0)$ to z at time t_1.

The maximizing λ for equation (8.8.4) was found numerically.

The minimum norm control $u(\tau)$ was then determined using equation (8.8.5). The results are shown in Figure 8.1.

8.9 EXERCISES

(1) Consider how on appropriate spaces, norms might be defined in a meaningful way to measure product quality in:

(i) a rolling mill producing metal strip where the important variable is thickness variation along the strip length, measured from a desired nominal thickness;

(ii) similar to (i), but now the thickness variation along the length and across the width of the strip has to be taken into account;

(iii) a reactor in which the temperature and (single valued) composition are required to follow specified constant spatial curves;

(iv) a furnace that is required to heat a load to a final temperature x_d with due consideration to the fuel u consumed.

(2) A rocket is required to shoot down an aircraft on a known track $x_d(t) \in \mathbb{R}^3$. The aircraft will be shot down if for any t the position $x(t)$ of the rocket is within a given distance r of the aircraft. Can you express this as a minimum norm problem? If not, consider which minimum norm formulation might be a good engineering approximation.

(3) Let a system described by the equations

$$\dot{x} = Ax + Bu, \qquad X = \mathbb{R}^2, u \in U,$$

U a normed space, require steering from an initial state x_0 to a given state x_d in a fixed time T while minimizing

$$J = \left(\int_0^T \|u\|^p \, dt \right)^{1/p}.$$

Let

$$A = \begin{pmatrix} -1 & 0 \\ 0 & -2 \end{pmatrix}, \qquad x_0 = \begin{pmatrix} 5 \\ 10 \end{pmatrix}, \qquad x_d = \begin{pmatrix} -10 \\ 5 \end{pmatrix}, \qquad T = 1 \text{ second.}$$

Calculate the vectors $\Phi(T)x_0$ and $x_d - \Phi(T)x_0$ and sketch these vectors and x_d in the $X(T)$ space. Convince yourself that the solution sought is the minimum norm pre-image of $x_d - \Phi(T)x_0$.

Under what conditions will such a pre-image exist? Under what conditions will it be unique? Set up an expression to characterize the required u function.

(4) Work through the proof that the minimum norm solution to drive the system

$$\dot{x}(t) = A(t)x(t) + B(t)u(t), \qquad x \in \mathbb{R}^n,$$

from $x_0 = x(0)$ to x_d in time T, $u \in U$ where U is the Hilbert space $(L^2[0, \infty))^r$ and the J to be minimized

$$J = \sum_{\lambda=1}^{r} \int_0^T |u_i(\tau)|^2 \, d\tau$$

is given by

$$u_{opt}(t) = B^T(t)\Phi^T(T, t) \left[\int_0^T \Phi(T, \tau)B(\tau)B^T(\tau)\Phi^T(T, \tau)\, d\tau \right]^{-1}$$

$$[x_d - \Phi(T, 0)x_0]$$

(i) Will the matrix inverse in the expression always exist?

(ii) This expression can be manipulated into a feedback law so that

$$u_{opt}(t) = g(x(t)).$$

Perform this manipulation.

(5) The first order system described by $a\dot{x}(t) = u(t)$ is to be driven from an initial state $x(0) = 0$ to a final state $x(T) = x_d$, T being fixed, while minimizing

$$J = \int_0^T u(t)^2\, dt$$

Show that a unique optimum control u_{opt} exists and that $u_{opt} = ax_d/T$. Notice that u_{opt} is a constant and that the solution represents an open loop control strategy.

(6) The first order system described by $\dot{x}(t) = x(t) + u(t)$ is to be driven from an initial state $x(0) = x_0$ to a final state $x(T) = 0$, T being fixed, while minimizing

$$J = \int_0^T (\gamma x(t)^2 + u(t)^2)\, dt$$

where γ is a constant.

Determine an expression for the control that achieves this optimal transfer.

(7) Define $X = C^0[a, b]$ as the set of functions f, continuous on $[a, b]$ and satisfying $f(a) = 0$. Define

$$\|f\| = \sup_{t \in [a, b]} \{|f(t)|\}$$

This is indeed a norm on X and X is a Banach space. Define

$$Tf = \int_a^b f(t)\, dt$$

Show that the problem; determine an f on $[a, b]$ such that $Tf = \alpha \in K$, $\alpha > 0$ given, while $\|f\|$ is minimized, has no solution. Consider the reasons why a solution does not exist.

(8) Let Σ be an nth order system that is to be driven to some target in minimum time. Show how this problem can be presented as a minimum norm problem.

    (*Hint*: Introduce an additional state variable.)

(9) Let X be a Hilbert space and $\{e_i\}$, $i = 1, \ldots, n$, be a set of orthonormal elements of X. Let E be the subspace generated by $\{e_i\}$. Let $x \in X$ be chosen arbitrarily. The element $y_0 \in E$ that satisfies

$$\|y_0 - x\| \langle \|y - x\|, \qquad y \in E, \quad y \neq y_0$$

is given by

$$y_0 = \sum_{i=1}^{n} \langle x, e_i \rangle e_i$$

Thus the best approximation, in a minimum norm sense, to an arbitrary element in a Hilbert space by an element restricted to a subspace is given by a Fourier series. y_0 can be considered geometrically as the projection of x onto the subspace E.

    Investigate in detail and relate the minimum norm control problem to the more general problem of best approximation in a least squares sense. (See Problem 8, Section 3.6.)

CHAPTER 9

Minimum Time Control

9.1 PRELIMINARIES AND PROBLEM DESCRIPTION

9.1.1 Introduction

Minimum time control is essentially concerned with driving a system to a desired state in minimum time. Such problems are physically meaningful only if constraints are imposed on the control variables u, since otherwise, trivially, the desired state could be achieved in zero time by the application of controls of infinite amplitude. For minimum time control, the set Ω of admissible controls becomes significant.

9.1.2 Problem description

Let Σ be a controllable continuous time system

$$\Sigma = \{I, U, \Omega, X, Y, \phi, \eta\}$$

where

$$U = L^{\infty}(I)$$

(this being the most appropriate setting for a minimum-time problem).

$$\Omega = \{u \,|\, \|u\| \leqslant k\}, \qquad k \text{ a given real number}$$

$$t_0 \in I, \qquad x(t_0) \in X, z(t) \in X$$

are given where z is continuous on I.

The minimum time problem is then: Determine $u(t)$ on $\Omega \times [t_0, t(k)]$ such that:

    (i) $x(t(k)) = z(t(k))$.
    (ii) $t(k) = \inf\{t \,|\, x(t) = z(t), t \in [t_0, \infty)\}$.

9.1.3 Comparison with the minimum norm problem

Consider the minimum time problem given in Section 9.1.2, and for the same system define a minimum norm problem as in Section 8.4.1.

Assume that t_1 for the minimum norm problem is defined so that $t_1 < t(k)$ then by definition there can be no solution to the minimum norm problem since $t(k)$ is the minimum time for which $x(t) = z$. A control u on $[t_0, t_1]$ such that $x(t_1) = z$ would satisfy the inequality $\|u\| > k$ and would not be a member of the set of admissible controls. However, for $t_1 > t(k)$ the minimum norm is well posed and is unaffected by the restriction of u to belong to the admissible set.

The case $t_1 = t(k)$ is of interest. From the definition of $t(k)$ a control u on $[t_0, t_k]$ exists such that $x(t(k)) = z$, while $\|u\| \leqslant k$. Intuitively, it can be expected that $\|u\| = k$.

The two problems are closely related. In the minimum norm problem $t_1 \in I$ is given and the value of $\|u\|$ on $[t_0, t_1]$ is free. In the minimum time problem the maximum value of $\|u\|$ is given while the time $t(k)$ is free.

9.1.4 Literature

Amongst an extensive literature the following are noteworthy:

Expository papers by La Salle (L2, L3, L4), Kreindler (K17), and the book by Hermes and La Salle (H8).

Papers concerned primarily with existence of time optimal controls Fillipov (F3), Roxin (R4), Markus and Lee (M1) and Schmaedeke (S3).

The paper by Kranc and Sarachik (K13) which takes an algorithmic approach to the characterization of time optimal controls.

Papers by Krasowskii (K14, K15), which were amongst the first to use functional analytic methods.

The attainable set and its properties play a key role in the theory of time optimal control. The papers by Halmos (H5) and Hermes (H7) are concerned with the derivation of these properties. Both papers make use of the theorem of Lyapunov on the range of a vector measure.

9.2 THE ATTAINABLE SET

9.2.1 Definition

Until further notice we assume that $X = \mathbb{R}^1$, the results obtained being easily extended to the case $X = \mathbb{R}^n$.

Let

$$S : \Omega \times I \to X$$

represent the mapping ϕ with t_0. $x(t_0)$ fixed. Clearly S is a continuous linear functional. The *attainable set* $\mathscr{A}(t)$ is defined by the relation

$$\mathscr{A}(t) = \{x(t) | x(t) = Su(t), u(t) \in \Omega \times I\}$$

Clearly if the system Σ is controllable we must have $z(t) \in \mathscr{A}(t)$ for some $t \in I$.

9.2.2 Properties of the attainable set

Theorem 9.1. *The attainable set is convex and compact.*

Proof. $\mathscr{A}(t)$ is the image under a linear mapping of the convex set Ω hence $\mathscr{A}(t)$ is convex. By the Banach–Alaoglu theorem (5.7), Ω, which is a closed ball in $L^\infty(I)$ is weak* compact.

The mapping Su can be represented

$$Su = \int_I g(t)u(t) \; dt$$

for some $g(t) \in L^1(I)$.

Since S is a functional defined on $L^\infty(I)$, S is weakly continuous. To complete the proof, S must be shown to be weak* continuous. The open sets in the weak* topology for $L^\infty(I)$ are of the form

$$\{u | g(u)| < d\} \quad \begin{array}{l} g(u) \text{ a linear functional on } L^\infty(I), \\ d \text{ a real number.} \end{array}$$

Let M be any open set in $\mathbb{R}^1$. For weak* continuity of S, the set $S^{-1} M$ is required to be open in the weak* topology of $L^\infty(I)$,

$$S^{-1}M = \{u | Su \in M\} = \{u | g(u) \in M\}$$

Assume that d is the diameter of the set M then

$$S^{-1}M = \{u| \, |g(u)| < d\}$$

and $S^{-1}M$ is open in the weak* topology for $L^\infty(I)$.

If M is the finite union of open sets in $\mathbb{R}^1$ then the above argument can be applied to each of the open sets. The inverse of M will again be the union of sets that are open in the weak* topology for $L^\infty(I)$. $\mathscr{A}(t)$ is a weak* continuous mapping of a weak* compact set Ω into $\mathbb{R}^1$ hence $\mathscr{A}(t)$ is compact in $\mathbb{R}^1$.

9.3 EXISTENCE OF A MINIMUM TIME CONTROL

Theorem 9.2. *There exists at least one time optimal control satisfying the problem conditions (i), (ii).*

Proof. From the controllability assumption,

$$z(t) \in \mathcal{A}(t) \quad \text{for some } t \in I$$

From the problem definition,

$$z(t) \in \mathcal{A}(t) \quad \text{for } t > t(k)$$

Let $\{t_n\}$ be a decreasing sequence in I satisfying

$$\{t_n\} \to t(k)$$

Let $\{u_n\} \to u$ be a sequence in Ω such that

$$S(t_n, u_n) = z(t_n)$$

Then by the triangle inequality

$$
\begin{aligned}
\left| z(t(k)) - x(t(k), u_n) \right| \\
\leqslant \left| z(t(k)) - z(t_n) \right| + \left| x(t(k), u_n) - x(t_n, u_n) \right| \\
\leqslant \left| z(t(k)) - z(t_n) \right| + \left| S(t(k), u_n) - S(t_n, u_n) \right|
\end{aligned}
$$

From the continuity of z and S, u is an optimal control and $t(k)$ is the minimum time.

9.4 UNIQUENESS

Theorem 9.3. *The optimal control of Theorem 9.2 is unique if and only if $z(t(k))$ is an extreme point of $\mathcal{A}(t(k))$.*

Proof. For brevity let $z = z(t(k))$. Let $t_0 = 0$ and let u on $[0, t(k)]$ be the unique optimal control. Assume that z is not an extreme point of $\mathcal{A}(t(k))$ then there exist points z_1, z_2 in $\mathcal{A}(t(k))$ such that

$$z = \tfrac{1}{2}z_1 + \tfrac{1}{2}z_2.$$

Define

$$y = z/2, \qquad y_1 = z_1/2, \qquad y_2 = z_2/2.$$

By the definition of $\mathcal{A}(t(k))$, z, z_1, z_2 each satisfy the relations

$$Su = z, \qquad Su_1 = z_1, \qquad Su_2 = z_2.$$

There must exist a real number $\alpha < 1$ such that

$$y = \phi(x(0), 0, u, \alpha t(k))$$

$$y_1 = \phi(x(0), 0, u_1, \alpha t(k))$$

$$y_2 = \phi(x(0), 0, u_2, \alpha t(k))$$

i.e. y, y_1, y_2 can be reached from the origin in a time $\alpha t(k)$. Now if the time for the trajectory $z - y_1$ to be traversed is equal to $(1 - \alpha)t(k)$, then the optimal control u to reach z cannot be unique. Now the trajectory $z_2 - y_2$ can be traversed in time $(1 - \alpha)t(k)$. Thus the trajectory $(y_1 - x_0) + (z_2 - y_2)$ can be traversed in a time $t(k)$ and the end point of the trajectory is z. Since

$$y_1 - x_0 + z_2 - y_2 = y_1 - x_0 + 2y_2 - y_2 = z - x_0$$

z can be reached in time $t(k)$ by the application of control u_1 over the interval $[0, \alpha t(k)]$ followed by the application of control u_2 over the interval $[\alpha t(k), t(k)]$.

To prove the second half of the theorem assume that u_1, $u_2 \in \Omega$ are two optimal controls on $[0, t(k)]$ and assume that the set

$$E = \{t | u_1(t) \neq u_2(t), \ t \in [0, t(k)]\}$$

is of positive measure. From the assumptions

$$\phi(x(0), 0, u_1, t(k)) = \phi(x(0), 0, u_2, t(k)) = z$$

while from the uniqueness theorems for differential equations there exists t_1 such that

$$y_1 = \phi(x(0), 0, u_1, t_1) \neq \phi(x(0), 0, u_2, t_1) = y_2.$$

Define

$$q_1 = \phi(y_1, t_1, u_2, t(k))$$
$$q_2 = \phi(y_2, t_1, u_1, t(k))$$

Clearly $q_1, q_2 \in \mathscr{A}(t(k))$ since both have been obtained by applying admissible controls over the time interval $[0, t(k)]$.

$$q_1 = \phi(\phi(x(0), 0, u_1, t_1), t_1, u_2, t(k))$$
$$q_2 = \phi(\phi(x(0), 0, u_2, t_1), t_1, u_1, t(k))$$
$$\tfrac{1}{2}(q_1 + q_2) = \phi(\phi(x(0), 0, \tfrac{1}{2}(u_1 + u_2), t_1), t_1, \tfrac{1}{2}(u_1 + u_2), t(k))$$
$$= \phi(x(0), 0, \tfrac{1}{2}(u_1 + u_2), t(k)) = z.$$

Hence z is an interior point of $\mathscr{A}(t(k))$.

9.5 CHARACTERIZATION

9.5.1 Key theorems

Theorem 9.4. *A control $u \in \Omega$ that is optimal must satisfy the inequality*

$$\|u\| \leq k$$

with equality.

Proof. By the assumption of controllability, ϕ is surjective onto X. ϕ is continuous and linear; U, X are complete. Define a set E,

$$E = \{x|\phi(x(t_0), t_0, u, t(k)), u \in \Omega^o\}$$

By the open mapping theorem (Theorem 4.5),

$$E \subset \mathscr{A}^o(t(k))$$

i.e. the transition mapping ϕ maps the interior of the set of admissible controls into an open set that is contained in the interior of the admissible set. For u to be an optimal control on $[t_0, t(k)]$ such that

$$Su = x(t(k))$$

we must have

$$z(t(k)) \in \partial\mathscr{A}(t(k))$$

This follows from the fact that S is continuous in t and that $z(t) \notin \mathscr{A}(t)$ for $t < t(k)$ so that $z(t(k))$ must be a boundary point of the attainable set $\mathscr{A}(t(k))$. Thus the pre-image of $z(t(k))$ in Ω cannot belong to the interior Ω^o and hence must belong to the boundary $\partial\Omega$,

$$u \in \partial\Omega \Rightarrow \|u\| = k.$$

Comment

Notice that this theorem asserts that

$$\text{ess sup } |u| = k \quad \text{on } [t_0, t(k)]$$

The theorem does not assert that $|u| \equiv k$ on the whole interval $[t_0, t(k)]$.
    Define a set E by the relation

$$E = \{x|Su = x; |u(t)| = k, \forall t \in [t_0, t(k)]\}.$$

(Recall that Su is defined by

$$Su = \phi(x(t_0), t_0, u, t(k))$$

and that

$$\mathscr{A}(t(k)) = \{x|Su = x; |u(t)| \leqslant k, t \in [t_0, t(k)]\}.).$$

Theorem 9.5.

$$E = \mathscr{A}(t(k))$$

The proof of this theorem is outside the scope of this book. The reader is referred to La Salle, (L3). La Salle's proof uses Lyapunov's theorem on the range of a vector measure. Briefly Lyapunov's theorem states that the range of every vector measure is closed and that under a further non-degeneracy condition (here satisfied) the range of a vector measure is convex. Using this approach, La Salle shows that the set E is closed and convex. He then shows that E is dense in $\mathscr{A}$. Since $E \subset \mathscr{A}$, then the conclusion is that $E = \mathscr{A}$.

Comment

La Salle's proof of Theorem 9.5 makes no use of the convexity of the set Ω. This leads to speculation concerning Theorem 9.1 where convexity of Ω was assumed although it could apparently be dispensed with.

This facet of the problem has been pursued by Hermes (H7). Specifically, Hermes derives the minimum conditions that can be imposed on the admissible set Ω such that the attainable set $\mathscr{A}$ will still be compact and convex.

A related paper by Neustadt (N4) examines yet another aspect—given that an attainable set A is compact but not convex, to what extent is it still possible to prove the existence of optimal controls?

Implications of Theorem 9.5.

Consider the time optimal control of a system Σ. Recall that $U = L^\infty(I)$ and $\Omega \subset U$ is defined

$$\Omega = \{u|\ \|u\| \leqslant k\}$$

In control theory a control u that satisfies the relation

$$|u(t)| = k \quad \text{for almost all } t \in I$$

is called a *bang-bang control*.

Now since $E = \mathscr{A}(t(k))$ the following corollaries are immediate consequences of Theorem 9.5.

Corollary 9.5.1. *If there exists an optimal control $u \in \Omega$ for the system Σ such that $x(t(k)) = z$ then there also exists a second optimal control $u' \in \partial\Omega$ with $|u'(t)| = k$ for almost all $t \in [t_0, t(k)]$. This can be summarized: If there exists an optimal control there also exists a bang-bang optimal control.*

Corollary 9.5.2. *If there exists a unique optimal control u for the system Σ then u is necessarily a bang-bang control.*

Corollary 9.5.3. *A bang-bang control on an interval I is necessarily of the form*

$$u(t) = k \operatorname{sign}(f(t))$$

where $f(t)$ is some function that is nonzero on I except possibly on a set of measure zero.

9.5.2 Algorithmic aspects

The following argument can form the basis for the development of optimal control algorithms.

Let

$$x(t_0) = 0, \qquad z = z(t(k)) \in \partial \mathscr{A}(t(k))$$

Then there exists a hyperplane M supporting $\mathscr{A}(t(k))$ at z. M can be represented as the translation of the null space of some nonzero linear functional g on the space X:

$$M = \{x | g(x) = c\}, \qquad c \text{ a real number}$$

$g(x)$ is also representable in the form

$$\langle x, g \rangle \qquad \text{where} \quad g \in X$$

where the element g is a normal to the hyperplane M.

The point z satisfies the conditions

$$z \in \mathscr{A}(t(k)) \cap M$$

$$\langle z, g \rangle = \sup_{x} \langle x, g \rangle \tag{9.5.1}$$

In words, z is the farthest point in $\mathscr{A}(t(k))$ in the direction g. Now $x = Su$ and if $z = Su$ with $z \in \partial \mathscr{A}(t(k))$, then u is an optimal control on $[0, t(k)]$. Further, if z is an extreme point of $\mathscr{A}(t(k))$ then u is the unique optimal control.

From equation (9.5.1)

$$\langle z, g \rangle = \sup_{u} \langle Su, g \rangle \tag{9.5.2}$$

Kreindler (K17), Kranc and Sarachik (K13), and Hemming (H6) have used the above principle to develop constructive algorithms for the synthesis of time optimal controls. The basic approach is as follows.

Let

$$\Sigma = \{I, U, \Omega, X, Y, \phi, \eta\}$$

be a time invariant system. Σ can also be represented by the equation

$$\dot{x}(t) = Ax(t) + Bu(t)$$

or equivalently by the equation

$$x(t) = \Phi(t)x(t_0) + \int_{t_0}^{t} \Phi(t - \tau)B(\tau)u(\tau)\,d\tau$$

$$= \Phi(t)x(t_0) + \Phi(t) \int_{t_0}^{t} \Phi(-\tau)B(\tau)\,u(\tau)\,d\tau$$

Given a continuous function $z(t) \in X$: Let the control objective be to choose $u(t) \in \Omega \subset U$, with U an L^p space such that:

(i) $x(t_k) = z(t_k)$.
(ii) $t_k = \inf\{t|x(t) = z(t), t \geq t_0\}$

$$\Phi^{-1}(t)x(t) = x(t_0) + \int_{t_0}^{t} \Phi^{-1}(\tau)B(\tau)u(\tau)\,d\tau.$$

Define

$$e(t) = \Phi^{-1}(t)z(t) - x(t_0).$$

The objectives will be met if

$$e(t_k) = \int_{t_0}^{t_k} \Phi^{-1}(\tau)B(\tau)u(\tau)\,d\tau = \int_{t_0}^{t_k} Q(\tau)u(\tau)\,d\tau$$

where

$$Q(\tau) \triangleq \Phi^{-1}(\tau)B(\tau)$$

Define

$$r(t) = \int_{t_0}^{t} Q(\tau)u(\tau)\,d\tau, \qquad r(t) \in X.$$

The control objective is then to choose u such that $r(t_k) = e(t_k)$. Assume there exists a unique optimal control then necessarily

$$e(t_k) \in \partial\mathscr{A}(t_k) \cap M$$

where

$$\mathscr{A}(t_k) = \{r(t_k)|u \in \Omega \times [t_0, t_k]\}$$
$$M = \{x|g(x) = c\}$$

for some function g and for some constant c.

From equation (9.5.2) for optimality

$$\langle e(t_k), g \rangle = \sup_u \langle r(t_k), g \rangle$$

$$= \sup_u \int_{t_0}^{t_k} Q(\tau)u(\tau)g \, d\tau$$

$$\leqslant \left(\int_{t_0}^{t_k} |Q(\tau)g|^q \, d\tau \right)^{1/q} \|u\|_p$$

$$\leqslant k \left(\int_{t_0}^{t_k} |Q(\tau)g|^q \, d\tau \right)^{1/q}.$$

The condition for equality to be attained in the inequality chain is that

$$u(\tau) = \alpha |Q(\tau)g|^{q/p} \operatorname{sign}(Q(\tau)g) \qquad (9.5.3)$$

α is a constant to be determined and

$$1/q + 1/p = 1$$

When U is an L^∞ space, equation (9.5.3) reduces to

$$u(\tau) = \alpha \operatorname{sign}(Q(\tau)g).$$

From Theorem 9.4 $\|u\|_\infty = k$ for optimality, hence $\alpha = k$ to give

$$u(\tau) = k \operatorname{sign}(Q(\tau)g)$$

t_k and g have to be computed and the references quoted above suggest methods.

Mizukami (M5) has developed these ideas further to obtain an algorithm for the minimum time control problem where the state variables are subject to instantaneous amplitude constraints at specific discrete times.

9.6 THE PONTRYAGIN MAXIMUM PRINCIPLE

The system Σ described by the differential equation

$$\dot{x} = f(x, u), \qquad x \in \mathbb{R}^n, \ U = (L^2)^r, \quad \Omega \subset U$$

where Ω is a k cube in U, is to be controlled such that the expression

$$J = \sum_{i=1}^{n} c_i x_i(T)$$

is minimized (maximized).

Additional *adjoint variables* $\psi_1, \ldots, \psi_n$ are introduced and defined by

$$\dot{\psi}_i = -\frac{\partial H}{\partial x_i}, \qquad i = 1, \ldots, n$$

where H is the Hamiltonian of the system defined by

$$H = \sum_{i=1}^{n} \psi_i \dot{x}_i = \langle \psi, \dot{x} \rangle$$

The maximum principle states that J will be minimized (maximized) if u is chosen at all times so as to maximize (minimize) the Hamiltonian H. (We note that this is a necessary condition that an optimal control u must satisfy but that it is not a sufficient condition.)

To apply the Pontryagin principle to the time optimal control problem we define a dummy state variable x_{n+1} by the relation $\dot{x}_{n+1} = 1$, and by suitably defining the time over which integration is to proceed we can make $J = x_{n+1}$.

Consider the time optimal problem: Given a system

$$\dot{x} = Ax + Bu, \qquad x \in \mathbb{R}^n, \qquad u \in (L^p)^r, \qquad u \in \Omega$$

where Ω is the k cube in U, determine the control u_{opt} such that $J = x_n$ is minimized over $[0, T]$.

The Hamiltonian H is given by

$$H = \langle \psi, \dot{x} \rangle = \langle \psi, Ax + Bu \rangle$$
$$\dot{\psi} = -\psi^T A.$$

This is a linear differential equation with solution

$$\psi(t) = T(t)\psi(0)$$

where $T(t)$ is a transition matrix. Define $M = \sup_u \{H\}$. Then by Pontryagin's principle,

$$H(\psi, x, t, u_{\text{opt}}) = M \quad \text{for almost all } t.$$

Now

$$M = \langle \psi, Ax \rangle + \sup_u \langle \psi, Bu \rangle$$

so that u_{opt} must maximize $\langle \psi, Bu \rangle$ at all times.

Define the matrix $D = \psi(t)^T B$ and let $D_i(t)$ be the ith row of D then if the bang-bang theorem is applicable, u_{opt} must be of the form

$$u_{opt} = k \begin{pmatrix} \text{sign}(D_1(t)) \\ \text{sign}(D_2(t)) \\ \cdots \cdots \\ \text{sign}(D_r(t)) \end{pmatrix}$$

(To obtain numerical values for the matrix $D(t)$ often involves a considerable computation. Both the original equation and the adjoint equation, $\dot{\psi} = -\psi^T A$, have to be solved simultaneously and, whereas the initial condition for x is known, only the final values for ψ are known and the problem is called a two point boundary value problem. Analytical techniques will work only in the simplest of cases and numerical techniques will nearly always be needed.)

Consider the time optimal problem of Section 9.5.2. The Hamiltonian for the problem is given by

$$H = \langle \psi(t), (Ax(t) + Bu(t)) \rangle$$

where the adjoint variables ψ are defined by the relation

$$\frac{\partial \psi}{\partial t} = -\frac{\partial H}{\partial x} = -\psi(t)A \qquad (9.6.1)$$

$u(t)$ will be a minimum time control if it maximizes H almost everywhere on $[t_0, t_k]$. Hence

$$u(t) = k\,\text{sign}(\psi(t)B)$$

Equation (9.6.1) is satisfied by

$$\psi(t) = \lambda \Phi^{-1}(t)$$

where λ is arbitrary, leading to

$$u(t) = k\,\text{sign}(\lambda \Phi^{-1}(t)B) = k\,\text{sign}(\lambda Q(t))$$

Assume that u is unique, then $\lambda = g$ in equation (9.8.10) and the interrelation of the methods can be seen.

Geometric interpretation

We note that the equation $H = \langle \psi, \dot{x} \rangle = c$, where H is the Hamiltonian and c is a constant, represents a hyperplane. Let $\Pi(t)$ represent this hyperplane then $\psi(t)$ is normal to Π for all t.

It is clear that maximizing the Hamiltonian is equivalent geometrically to choosing u so that the projection of the vector $\dot{x}(t)$ on to $\psi(t)$ is maximized.

Now we can show that if $u_{\text{opt}}(t)$ is the time optimal control then

$$x(t, u_{\text{opt}}) \in \partial \mathcal{A}(t), \qquad \forall t \in [0, t_{\text{opt}}].$$

In words, $x(t, u_{\text{opt}})$ belongs to the boundary of the attainable set for every $t \leqslant t_{\text{opt}}$.

Let Π be a hyperplane through $x(t_{\text{opt}})$, then Π supports $\mathcal{A}(t)$. The normal to Π is $\psi(t)$ and the projection of the motion $\dot{x}$ onto $\psi(t)$ is given by

$$l = \left\langle \dot{x}, \frac{\psi(t)}{\|\psi(t)\|} \right\rangle \frac{\psi(t)}{\|\psi(t)\|}$$

Figures 9.1 and 9.2 show the geometric interpretation. The optimal trajectory passes through the boundary of each attainable set $\mathcal{A}(t)$ $t < t_{\text{opt}}$ such that the projection of the instantaneous motion onto the normal $\psi(t)$ is maximized.

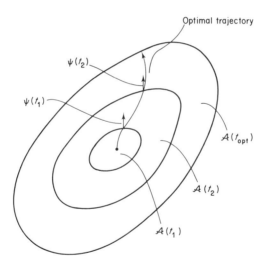

Figure 9.1.

Pontryagin shows that it is a necessary condition of optimal control that the trajectory obeys this geometric condition.

From this geometric viewpoint it can be seen that the optimal control of the form $u_{\text{opt}} = k\,\text{sign}(D(t))$ maximizes for every t the magnitude of the projection of the motion on to the normal to the attainable set $\mathcal{A}(t)$.

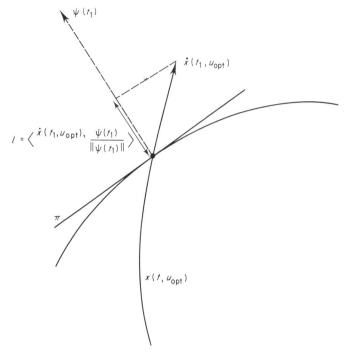

Figure 9.2. Figures 9.1, 9.2 show the relation between the time-optimal trajectory, the boundary of the attainable set and the adjoint variable.

9.7 TIME OPTIMAL CONTROL: DISCUSSION

(The statements and theorems in this section are not proved for reasons of space. Proofs can be found in Hermes (H8).)

Two questions that arise naturally in connection with time optimal control are:

(i) When is time optimal control unique and hence bang-bang?
(ii) Can the optimal control be determined algorithmically by application of the necessary conditions derived earlier?

Geometrically, the boundary of the attainable set $\mathscr{A}(t)$ plays a key role. If the set is rotund with a smooth boundary, no problems would be expected and none occur. The optimal control is unique and it can be determined from the necessary condition.

When $\partial\mathscr{A}(t)$ is not smooth then $\mathscr{A}(t)$ is a set with "corners". Through some boundary points of $\mathscr{A}(t)$ pass many hyperplanes of support. However, provided that $\mathscr{A}(t)$ is rotund, a unique time optimal control still exists and it can be calculated by taking any of the possible hyperplanes.

A more important difficulty arises when $\mathscr{A}(t)$ is not rotund. Under the assumptions made, $\mathscr{A}(t)$ must be convex so $\mathscr{A}(t)$ must have a "flat side". The same hyperplane of support passes through many boundary points of $\mathscr{A}(t)$. Under this condition, time optimal control is no longer unique, although, of course, there is, by earlier theorems, at least one bang-bang time optimal strategy for each boundary point of $\mathscr{A}(t)$. Nor can it be guaranteed that time optimal control be calculated from the necessary conditions. Thus, rotundity of $\mathscr{A}(t)$ guarantees favourable asnwers to the two questions posed at the start of this discussion—summarized in a theorem:

Theorem 9.6. *Let Σ be a system for which the set $\mathscr{A}(t)$ is rotund for all t. Then time optimal control for Σ is unique, bang-bang, and can be calculated from the necessary condition.*

The next question that arises is bound to be: Given a system described by the equation $\dot{x} = A(t)x + B(t)u$, under what conditions will $\mathscr{A}(t)$ be rotund? We do not pursue this general case but limit ourselves to the time invariant equation.

Theorem 9.7. *Let Σ be a time invariant system described by the equation $\dot{x} = Ax + Bu$, with $u \in \Omega$, Ω the k cube in U. Then $\mathscr{A}(t)$ is rotund for all t if and only if the set of vectors $\{b^j, Ab^j, \ldots, A^{n-1}b^j\}$ is linearly independent for $j = 1, \ldots, r$, where r is the dimension of the u vector.*

9.8 EXERCISES

(1) In the equation

$$\begin{pmatrix} \dot{x}_1 \\ \dot{x}_2 \end{pmatrix} = A\begin{pmatrix} x_1 \\ x_2 \end{pmatrix} + \begin{pmatrix} 0 \\ 1 \end{pmatrix}u$$

the matrix A has real negative eigenvalues. The system is to be driven from some initial condition

$$\begin{pmatrix} x_1 \\ x_2 \end{pmatrix}_0$$

to the origin in minimum time while u is subject to the constraint $|u(t)| \leq k$.

A bang-bang control for this system will be of the form

$$\begin{cases} u(t) = k, & 0 \leqslant t < \tau \\ u(t) = -k, & \tau \leqslant t < T \end{cases}$$

or

$$\begin{cases} u(t) = -k, & 0 \leqslant t < \tau \\ u(t) = k, & \tau \leqslant t < T \end{cases}$$

where T is the minimum time and τ is the time of control change-over. An effective practical algorithm for time optimal control can be developed by calculating solutions corresponding to $u = +k, u = -k$ in backwards time from the origin and using these solution curves as switching lines. Complete the details, sketching a block diagram of the resulting algorithm.

(2) Let a system be described by

$$\dot{x} = Ax + Bu, u \in \mathbb{R}^1, x(0) = 0$$

It is required to drive the system to $x_d \in X$ in minimum time while u is subject to the constraint $|u| \leqslant k$. Consider the problem from the viewpoint that the rate of change of x must be maximized in the direction x_d, i.e. that at all times $\langle x_d, \dot{x}(t) \rangle$ is to be maximized.
Show that this implies

$$u_{\text{opt}}(t) = k \operatorname{sign} \left(\langle x_d, \Phi^{-1}(t)B \rangle \right)$$

(3) (i) Prove that $\mathscr{A}(t)$ is convex if Ω is convex.
  (ii) Prove that $\mathscr{A}(t)$ is compact if Ω is weakly compact (when is Ω weakly compact?).

(4) Define X as the set of all compact sets in $\mathbb{R}^n$. Show that the mapping $\mathscr{A}(t):[0, \infty) \to X$ is continuous in t. (First define a metric on X.)

(5) Let Σ be a time invariant system for which the attainable set $\mathscr{A}(t)$ is rotund for all $t > 0$. Show that Σ is controllable. Not every controllable system has a rotund attainable set. Prove this by constructing a (time invariant) example.

(6) Show by calculation using the single-input-single-output system $\ddot{x} + x = u$ with $x(0) = 0$ that:
  (i) For $t < T$, the attainable set $\mathscr{A}(t)$ has boundary points where a hyperplane of support is not uniquely defined.
  (ii) For $t \geqslant T, \mathscr{A}(t)$ has a unique hyperplane of support at every boundary point.

Determine the value of T and comment on the implications. (This example is due to Hermes.)

(7) Given a system described by the equation $\dot{x} = Ax + Bu$, $x \in \mathbb{R}^n$, $U = (L^2)^r$, Ω the unit cube in U, which is to be driven from an initial state x_0 to a given final state x_d in minimum time. Make the substitution $q = \Phi(t)x$ and formulate the optimal control problem in terms of q, where it has a particularly simple form. Define the *reachable set* $\mathcal{R}(t)$ as the set of all q that can be reached in time t by using admissible controls over the interval $[0, t]$. Show by methods parallel to those for $\mathcal{A}(t)$, that the reachable set is convex and compact. The set $\mathcal{A}(t)$ is not symmetric. Is the reachable set symmetric?

(8) Let a single-input-single-output system be described by the equation $\ddot{x}(t) = u(t)$, $|u(t)| \leq M$, for all t. Determine the optimal control to drive the system from x_0 to the origin in minimum time.

(9) Given the equation

$$\dot{x}(t) = A(t)x(t) \tag{1}$$

The adjoint system is defined by

$$\dot{\psi}(t) = -A(t)^T \psi(t) \tag{2}$$

Determine the relation between the transition matrix of (1) and the transition matrix of (2).

CHAPTER 10

Distributed Systems

10.1 INTRODUCTION

The study of systems described by partial differential equations, here referred to as *distributed systems*, arises whenever the spatial distribution of variables is to be controlled (a typical application is the temperature profile in a reactor). The normal approach to distributed systems is either to introduce fictitious spatial discretization into the dynamic model or to introduce numerical discretization of the differential equation. The approach used here, based largely on work of Balakrishnan, considers the system state to be specified by one or more functions and attempts to preserve the continuous nature of the system. With this approach, the state space is infinite dimensional, a transition mapping can still be defined, and a theoretical development similar to that for systems described by ordinary differential equations can be obtained.

However, there are essential differences between the two cases:

(i) The theory of existence and uniqueness of solutions for partial differential equations is incomplete (compared with the situation for ordinary differential equations). In particular, no necessary conditions for existence and uniqueness of solutions of the evolution equation $\dot{x} = Ax$, where A is a partial differential operator, are known.

(ii) The differential operators associated with distributed systems are in general unbounded.

(iii) For distributed systems, the transition mapping no longer necessarily satisfies group axioms since an inverse transition mapping may not exist. (This is linked to the question of reversibility of dynamical processes). However, the transition mapping does satisfy the axioms of a transformation semi-group so that analytical tools such as the Hille–Yosida theorem can be brought to bear.

(iv) In the finite dimensional case, controllability and observability proofs make use of similarity transformations, equivalent to a change of basis in the state space. No equivalent similarity transformation is available for distributed systems.

Literature

Distributed parameter control problems have been studied extensively by Butkovsky (B14–B18). Balakrishnan (B1–B3) has produced a valuable body of basic theory on these systems and has pioneered the application of semi-groups to them. Friedman (F8) has established links from particular types of partial differential equations that allow the methods of Balakrishnan to be applied. Miranker (M4) has considered the controllability of distributed systems while Wang (W1–W6) has produced more practically oriented papers. A paper by Hullett (H12) demonstrates a practical application of some of the above ideas.

Other authors who have contributed to the study of these systems include Lions (L9–L10), Porter (P12), Kim (K7–K8), Datko (D2), Swiger (S10), Alvarado (A4) and Kisynski (K10).

10.2 FURTHER THEOREMS FROM FUNCTIONAL ANALYSIS

This chapter makes greater mathematical demands as has been explained in the introduction. Further supporting definitions and theorems are also required and these now follow. These are standard results in many texts on functional analysis.

10.2.1 Regulated functions

Let $f : I \to X$ be a mapping from an interval $I \subset \mathbb{R}^1$ to a Banach space X. At a point $x \in I$ we say that f has a *limit on the right* if

$$\lim_{\substack{y \in I, \, y > x \\ y \to x}} f(x)$$

exists. *Limit on the left* is defined similarly. These are called *one sided limits* of f. A mapping f of I into X is called a *regulated function* if it has a one sided limit at every point of I.

10.2.2 Semi-groups

Let E be a set of elements $a, b, \ldots$, with $\circ$ being a binary operation in E such that $a \circ b = c \Rightarrow c \in E$ then $\{E, \circ\}$ is called a *groupoid*.

A *semi-group* is a groupoid in which the binary operation is associative.

A *transformation semi-group* Γ is a realization of an abstract semi-group E if given elements $a, b \in E$ there exists elements $T(a), T(b) \in \Gamma$ such that:

(i) $T(a)T(b) \in \Gamma$.
(ii) $T(a \circ b) = T(a)T(b)$.

In general T can operate on n parameters $T(\alpha_1, \alpha_2, \ldots, \alpha_n)$ and Γ is said to be an *n parameter semi-group* of transformations.

(Notice that in this chapter we use the word transformation interchangeably with the word mapping, the two terms having the same meaning. Such usage makes for harmony with the literature.)

Let $\Gamma = \{T(a)|a > 0\}$ be a one parameter semi-group of linear bounded transformations on a complex Banach space X to itself such that

$$T(a_1 + a_2)x = T(a_1)(T(a_2)x), \qquad \forall a_1, a_2 > 0, \quad \forall x \in X$$

$T(a)$ is assumed strongly continuous for $a > 0$.

The *infinitesimal operator* A_0 of Γ is defined as the limit in norm as $h \to 0^+$ of

$$A_h x = (T(h) - I)x/h$$

A_0 is an unbounded linear operator whose domain is dense in the union of the range spaces of $\{T(\alpha)|\alpha > 0\}$. A_0 is not in general a closed operator, but the least closure of A_0 is called the *infinitesimal generator* of the semi-group. The set of elements for which $\lim_{h \to 0^+} A_h x$ exists is defined as the domain of A_0 and written as $D(A_0)$.

If $\lim_{a \to 0} T(a)$ exists then necessarily the limit must be the identity operator. If this limit exists in the uniform operator topology an infinitesimal generator A exists, A is bounded and $T(\zeta) = \exp(\zeta A)$. In fact this only occurs in the finite dimensional case when $\exp(\zeta A)$ is simply the transition matrix. In case the limit exists in the strong operator topology, a unique infinitesimal generator A still exists but A is now an unbounded transformation and the function $\exp(\zeta A)$ is not necessarily meaningful.

The type of convergence with which the limit is approached is important. Hille and Phillips list six different types of convergence.

However, if $T(h)x \to x$ as $h \to 0^+$ then $T(\zeta)x$ can be represented by the expression

$$T(\zeta)x = \lim_{h \to 0^+} \exp(\zeta A_h)x, \qquad \forall x \in X, \forall \zeta > 0.$$

(This expression being equivalent to the expression

$$T(\zeta) = \exp(\zeta A)$$

for the case where T is uniformly continuous.)

10.2.3 Theorems on semi-groups

Theorem 10.1. *Let A be the infinitesimal generator of a strongly continuous semi-group Γ of transformations*

$$\Gamma = \{T(t)|t \geq 0\} \qquad \text{with} \quad T(t), t \geq 0$$

mapping a complex Banach space E into itself, then the domain of A is dense in E.

Theorem 10.2. *The infinitesimal generator of a strongly continuous semi-group is closed.*

Theorem 10.3. *Given a linear operator A with domain $D(A)$. Let $x(0) \in D(A)$ and let $\{T(t)|t \geq 0\}$ be a semi-group of transformations, then:*

(i) $T(t)x(0) \in D(A)$.

(ii) $\dfrac{\mathrm{d}}{\mathrm{d}t}\left(T(t)x(0)\right) = AT(t)x(0)$.

Proof. Define A_h by the relation

$$A_h x = \frac{1}{h}\left(T(h)x - x\right)$$

such that

$$Ax = \lim_{h \to 0} A_h x$$

Choose $t \geq 0, h > 0$,

$$T(t)A_h x(0) = T(t)\frac{1}{h}\left(T(h)x(0) - x(0)\right)$$

$$= \frac{1}{h}(T(t + h)x(0) - T(t)x(0))$$

$$= A_h T(t)x(0)$$

Thus $T(t)$ and A_h commute.

$$\lim_{h \to 0} A_h T(t)x(0) = \lim_{h \to 0} T(t)A_h x(0)$$

and

$$T(t)x(0) \in D(A)$$

Also

$$AT(t)x(0) = \lim_{h \to 0} A_h T(t)x(0) = T(t)Ax(0)$$

$$\frac{1}{h}\left(T(t + h)x(0) - T(t)x(0)\right) = T(t)A_h x(0)$$

in the limit as $h \to 0$.

$$\frac{d}{dt}\,(T(t)x(0)) = T(t)Ax(0) = AT(t)x(0)$$

Theorem 10.4. (The Hille–Yosida theorem). (This is not the most general statement of the theorem.) *Define the* resolvent $R(s, A)$ *of the operator A by $R(s, A) = (sI - A)^{-1}$. Let A be a closed linear operator with dense domain, then a necessary and sufficient condition that A be the infinitesimal generator of a strongly continuous semi-group Γ is that for some real number k, all $s > k$ are in the resolvent set of A and*

$$\| R(s, A) \| \leqslant (s - k)^{-1}$$

for all real $s > k$.

The resolvent set and the spectrum of an operator

Let A be a linear operator $A: X \to X$ where X is a complex Banach space. The set of all s such that Range $(sI - A)$ is dense in X and such that $(sI - A)$ has a continuous inverse defined on its range is said to be the *resolvent set* of the operator A, denoted by $\rho(A)$. The set of all complex numbers s not in the resolvent set is defined to be the *spectrum* of the operator A, denoted by $\sigma(A)$.

10.3 AXIOMATIC DESCRIPTION

The axioms defined in Chapter 6 are applicable to the description of linear distributed dynamic systems and they are used unmodified in this chapter. In fact, a system description where the state space is a Banach space is one of the most general descriptions possible for a linear dynamic system, even allowing stochastic problems to be represented. However, the following points should be noted. In much of the literature, see for instance Wang (W2), additional structure is added to the axiomatic description for distributed systems so that there is a state space of spatial vectors and a state function space whose functions are defined on the state space. This type of formulation makes somewhat easier the link between a physical model and its description. For instance, in considering problems with moving physical boundaries, as in the solidification of molten metals, the varying spatial domain can be represented explicitly in the model.

10.4 REPRESENTATION OF DISTRIBUTED SYSTEMS

10.4.1 Representation of distributed systems in the form $\dot{x} = Ax + Bu$

Consider a partial differential equation such as the diffusion equation

$$\frac{\partial x(t, z)}{\partial t} = \frac{\partial^2 x(t, z)}{\partial z^2}, \qquad x(0, z) = f(z)$$

Introduce an operator $A : X \to X$ defined by

$$Ax(t, z) = \frac{\partial^2 x}{\partial z^2}(t, z)$$

so that

$$\frac{\partial x(t, z)}{\partial t} = Ax(t, z)$$

$$\dot{x}(t)(z) = Ax(t, z)$$

which is of the general form

$$\dot{x} = Ax \tag{10.4.1}$$

and this equation is a direct generalization of the finite dimensional case. A solution of equation (10.4.1) is now sought in the form

$$x(t) = G(t, t_0)x(0) \tag{10.4.2}$$

where $G(t, t_0) : X \to X$ is a linear operator for all t, t_0 with $t \geqslant t_0$.

In order that equation (10.4.2) shall represent the time solution of a dynamic system, G must satisfy the conditions

(i) $G(t_0, t_0) = I$.
(ii) $G(t_2, t_1)G(t_1, t_0) = G(t_2, t_0)$.

In case the system Σ is time invariant, G can be replaced by a transformation $T(t - t_0)$ which is a function of an interval in I. Thus

$$T(t_2 - t_1)T(t_1 - t_0) = T(t_2 - t_0), \qquad t_2 \geqslant t_1 \geqslant t_0$$

Let

$$t = t_2 - t_1, \qquad s = t_1 - t_0$$

then

$$t_2 - t_0 = s + t \geqslant 0$$

Hence

(i) $T(t)T(s) = T(t + s)$.
(ii) $T(0) = I$.

Consider the set $\{T(t)|t \geqslant 0\}$ with the binary operation in (i) above. The operation is associative and $\{T(t)|t \geqslant 0\}$ with the operation forms a transformation semi-group.

From the system axioms the response $x(t) = T(t - t_0)x(t_0)$ is required to be continuous in $x(t_0)$. This requires that $T(t) \in B(X)$ for all $t \geqslant t_0$ (or in general for all $t > 0$). A further continuity requirement imposed by the axioms for dynamic systems is that $T(t)x$ is continuous in t for all $t \in [0, \infty)$. This condition will be met if

$$\lim_{t \to t_0^+} \| T(t)x - T(t_0)x \| = 0$$

This is precisely the condition for strong continuity of T. The situation can be summarized. In order to be consistent with the axioms for a linear time invariant dynamic system, $\{T(t)|t > 0\}$ must form a strongly continuous semi-group of transformations.

Two questions now arise:
(i) What are the conditions on the operator A of equations such as (10.4.1) that guarantee the existence of a strongly continuous semi-group of transformations T? (This is considered in Section 10.4.2.)
(ii) How can T be characterized in terms of A?

10.4.2 Conditions on the operator A

Referring to Section 10.2.2, corresponding to the strongly continuous semi-group Γ there exists a unique (in general unbounded) infinitesimal generator that can be identified with the operator A of equation (10.4.1). A must therefore be the infinitesimal generator of a strongly continuous semi-group Γ if equation (10.4.1) is to satisfy the axioms for a dynamic system. From Section 10.2.3 two necessary conditions on A for this to be achieved are:

(i) A must be a closed operator.
(ii) The domain $D(A)$ of A must be dense in X.

Summary

The equation $\dot{x} = Ax$ with initial condition $x(0)$ has the solution

$$x(t) = T(t)x(0)$$

Provided that

$$x(0) \in D(A)$$

Notice that if the equation $\dot{x} = Ax$ is to represent a dynamic system, then A must be the infinitesimal generator of a strongly continuous semi-group. From Theorem 10.1 the domain $D(A)$ must be dense in X. This fact is fundamental in the proper formulation of distributed system models.

The second necessary condition on A is (from Theorem 10.2) that A must be a closed operator.

10.4.3 Uniqueness of solution

Theorem 10.5. *Let X be a Banach space and let A be a closed linear operator $A:X \to X$ with domain $D(A)$ dense in X.*

Let A be the infinitesimal generator of a strongly continuous semi-group

$$\Gamma = \{T(t)|t \geqslant 0\}$$

then the equation

$$\dot{x} = Ax \qquad x(0) = x_0$$

has the unique solution

$$x(t) = T(t)x_0, \qquad t \geqslant 0$$

provided that $x_0 \in D(A)$.

Proof. From Theorem 10.3,

$$\frac{d}{dt}(T(t)x_0) = A(T(t)x_0)$$

Let x^1, x^2 be two solutions such that $x^1(0) = x^2(0)$. Put $z = x^1 - x^2$, then $\dot{z} = Az$ with $z(0) = 0$. Then $z(t) \equiv 0$ for all $t \geqslant 0$, and uniqueness follows.

The requirement that $\{T(t)|t \geqslant 0\}$ be a semi-group is also necessary for uniqueness, see Phelps (P3).

10.4.4 Possible use of the Schauder base in system representation

Consider the solution of equation (10.4.1) with t fixed then $x(z) \in X$, where X is a Banach space. If there exists a Schauder base $\gamma(z)$ for X, then there exists a unique sequence of scalars $(\alpha_1, \alpha_2, \ldots)$, corresponding to $x(z)$ such that

$$\Sigma \, \alpha_i \gamma_i(z) \to x(z) \quad \text{strongly.}$$

In case X is the Hilbert space $L^2(z)$ then the system state can be specified by a unique sequence of Fourier coefficients in $l^2(z)$.

10.4.5 Distributed systems with distributed control

Consider the equation

$$\dot{x} = Ax + Bu \tag{10.4.3}$$

where $x \in X$, X a Banach space;

A is a linear unbounded operator with $D(A) \subset X$;

$u \in U$, where U is a Banach space;

B is a linear mapping $B : U \to X$.

As before, we require that A is the infinitesimal generator of a strongly continuous semi-group.

Let Bu be a regulated function of the variable t for $t \geqslant t_0$, for all $u \in U$. Let $Bu \in D(A)$. Let

$$\int_{t_0}^{t} \| ABu \| \, dt < \infty$$

Then the unique solution of equation (10.4.3) is

$$x(t) = T(t - t_0)x_0 + \int_{t_0}^{t} T(t - \tau)Bu(\tau) \, d\tau \tag{10.4.4}$$

(Note carefully that the integrand takes values in an infinite dimensional space and that Lebesgue integration is not applicable. The integral in equation (10.4.4) is a Bochner integral, this being a suitable extension of the Lebesgue integral. Bochner integration is fully discussed in Yosida (Y1, Chapter 5).)

A proof can be seen in Balakrishnan (B1).

10.5 CHARACTERIZATION OF THE SOLUTION OF THE EQUATION $\dot{x} = Ax + Bu$

Recall that the resolvent of the operator A is defined

$$R(s, A) = (sI - A)^{-1}, \qquad s \text{ a complex number.}$$

By analogy with the scalar case where

$$\mathscr{L}(e^{\gamma t}) = (s - \gamma)^{-1} = R(s, \gamma)$$

and with the case where E is a finite matrix where

$$\mathscr{L}\{e^{Et}\} = (sI - E)^{-1} = R(s, E)$$

($\mathscr{L}$ indicating Laplace transformation) it may be expected that $T(t)$ can be characterized in terms of the inverse Laplace transform of the resolvent.

For time invariant distributed systems it can be shown (see reference B2) that

$$T(t) = \mathcal{L}^{-1}(R(s, A)) = \mathcal{L}^{-1}((sI - A)^{-1})$$

and since the unique solution to the time invariant equation

$$\dot{x} = Ax + Bu$$

is given by

$$x(t) = T(t - t_0)x_0 + \int_{t_0}^{t} T(t - \tau)Bu(\tau) \, d\tau$$

$$x_0 \in D(A)$$

then inverse Laplace transformation of the resolvent can lead to a solution of the system equations.

This approach is exactly analogous to one method of calculating transition matrices for finite state systems although for distributed systems the approach may yield numerically intractable equations.

10.6 STABILITY

Stability in the sense of Lyapunov and asymptotic stability are defined for distributed systems in terms of the transition mapping exactly as for finite dimensional state systems in Section 7.2.2. However, the actual determination of stability is more difficult than for systems with finite dimensional state.

Extensions of the second method of Lyapunov to distributed systems have been undertaken by Zubov and by Massera. Zubov defines Lyapunov functionals on the infinite dimensional state space, whereas Massera extends Lyapunov's method to systems described by denumerably many ordinary differential equations. Both approaches are summarized in Section (III) of reference (W2).

More in keeping with the spirit of the present work are the following theorems.

Theorem 10.6. Let Σ be a time invariant distributed system. Let the system operator A be the infinitesimal generator of a semi-group and let there exist a positive constant k such that

$$\mathcal{R}(s) \leqslant -k, \qquad \forall s \in \sigma(A)$$

Then the system Σ is asymptotically stable to the origin of the state space.

Proof. From the Hille–Yosida theorem (Theorem 10.4),

$$\|e^{tA}\| \leqslant e^{-kt}, \qquad \forall t \geqslant 0$$

Since $x(t) = e^{(tA)}x(0)$ is the zero input solution of the system Σ, the theorem follows.

Theorem 10.7. *Let Σ be a time invariant distributed system and let the system operator A be the infinitesimal generator of a semi-group. Let the spectrum of A satisfy*

$$\mathscr{R}(s) \leqslant 0, \qquad \forall s \in \sigma(A)$$

Then the system is stable in the sense of Lyapunov.

Proof. From the Hille–Yosida theorem (Theorem 10.4),

$$\|e^{(tA)}\| \leqslant e^0 \qquad \text{for all } t \geqslant 0$$

Thus

$$\|e^{(tA)}\| \leqslant 1 \qquad \text{for all } t \geqslant 0$$

and the system is stable in the sense of Lyapunov.

Comment

In the two theorems above, the condition that A be the infinitesimal generator of a semi-group will follow automatically if the system satisfies the axioms for a dynamic system. Stability determination consists of obtaining bounds for the real part of the spectrum of A. No general methods appear to exist for this but particular systems have been studied. See for example Wang (W2).

10.7 CONTROLLABILITY

Let Σ be a dynamical system. Let X, the state space, be infinite dimensional. The controllability definitions are exactly the same as those for a system with finite dimensional state.

Thus, a system Σ is defined to be *controllable at t_0*, *completely controllable at t_0*, or *completely controllable for every t_0*, according as it satisfies the definitions of Section 7.3.1.

Although these definitions of controllability are nominally applicable to distributed systems, controllability of a particular system is heavily dependent on the problem formulation, e.g. on the condition imposed upon the input and state spaces. Further, there are no counterparts to the tests for controllability available for the finite dimensional case. Finally, controllability as defined above is too strong a requirement to impose in general on distributed systems.

Accordingly, for distributed systems, a weaker property than controllability is defined. Broadly a system is defined to be *weakly controllable* if, given any element $x_1 \in X$, there exists an admissible control such that the system state x can be brought arbitrarily close to x_1 in finite time. This concept is formalized in the following definition.

Definition. *The system Σ is defined to be weakly controllable at time t_0 if given an arbitrary element $z \in X$ and an arbitrary real number $\varepsilon > 0$, an admissible control $u \in \Omega$ exists such that*

$$\| \phi(0, t_0, u, t) - z \| < \varepsilon$$

for some finite $t \in I, t > t_0$.

If the system Σ is weakly controllable at all $t_0 \in I$ then Σ is defined to be simply *weakly controllable*.

Theorem 10.8. *Let Σ be a time invariant distributed system and let the state space X be a Hilbert space.* Let the zero state response of the system be written

$$x(t) = H(t_0, t)u$$

A necessary and sufficient condition for weak controllability at t_0 of the system Σ is that the adjoint mapping H^* be one to one.

Proof. (Necessity)

From the assumption of weak controllability, given $z \in X$ and a real $\varepsilon > 0$, there exists $u \in U$ and $t \in I$ such that

$$\| H(t_0, t)u - z \| < \varepsilon$$

Assume that $H(t_0, t)^* z = 0$. If it follows that $z = 0$ then H^* must be one to one. Then for all $u \in U$ (from the definition of adjoint),

$$\langle u, H(t_0, t)^* z \rangle = 0 = \langle H(t_0, t)u, z \rangle$$

From the above

$$\| H(t_0, t)u - z \|^2 < \varepsilon^2$$

Hence

$$\langle H(t_0, t)u - z, H(t_0, t)u - z \rangle = \langle H(t_0, t)u, H(t_0, t)u \rangle + \langle z, z \rangle$$
$$= \| H(t_0, t)u \|^2 + \| z \|^2 < \varepsilon^2$$

Thus $\|z\| < \varepsilon$ and since ε is arbitrary, $z = 0$, i.e.

$$H(t_0, t)^*x = 0 \Rightarrow x = 0$$

and H^* is a one-to-one mapping, as was to be shown.

Sufficiency

Define a set E by the relation

$$E = \{x \mid x = H(t_0, t)u; \forall t \in I, t \geqslant t_0, \forall u \in U\}$$

Then the system Σ will be weakly controllable at t_0, if $\bar{E} = X$, since elements in X can be approximated closely by elements of $H(t_0, t)u$. From Lemma 8 of Section (VI.2.6) in reference D12,

$$\bar{E} = \{x \mid \langle y, x \rangle = 0 \qquad \text{for all } y \in X \text{ with } H^*y = 0\}$$

Now if H^* is one to one then

$$H^*y = 0 \Rightarrow y = 0$$

Thus

$$\bar{E} = \{x \mid \langle 0, x \rangle = 0\} = X$$

Observability

As for finite state systems observability will be only briefly discussed. First, it is pointed out that the duality between controllability and observability established by Kalman for finite state systems no longer exists in general for distributed systems. Second, the observability of distributed systems is connected to the question of whether the semi-group of transformations of the system actually possesses the properties of a group. The best introductory reference on observability of distributed systems is possibly Wang (W2).

10.8 MINIMUM NORM CONTROL

A generalized minimum norm problem for distributed systems is the following.

Given a system Σ with input space U and state space X, both being real Banach spaces, and given $t_0, t_1 \in I$, $x(t_0)$, $z \in X$ determine $u \in U$ such that:

   (i) $x(t_1) = z$.

   (ii) $\|u\|$ is minimal.

u is then called a minimum norm control.

Define by S the transition mapping, restricted as shown below

$$Su = \phi(x(t_0), t_0, u, t_1)$$

For the finite state minimum norm problem, S was either a linear functional or a mapping of finite rank, representable in terms of a finite sum of functionals. For distributed systems we cannot make the assumption that Range $S_{t_1 \to \infty} = X$. However, if the system Σ is weakly controllable we can assume that Range S is dense in a subspace of X while Range $S_{t_1 \to \infty}$ is dense in X. Porter (P12) shows that if the space U is reflexive then for every $x \in$ Range (S) a minimum norm pre-image exists. The proof rests on the condition, ensured by reflexivity of the space U, that the image under S of the unit ball B of U is closed, and that the support points of a closed convex set in a Banach space are dense in its boundary. Thus for a weakly controllable distributed system whose input space is reflexive, a minimum norm control u exists that forces the system arbitrarily closely to any desired state z. Further, assuming still that Range (S) is dense in X then if the space U is rotund and smooth, the minimum norm element is unique. The proof of uniqueness is similar to that given in Chapter 8.

Characterization of the minimum norm control

Swiger (S10) has considered the characterization of minimum norm controls for distributed systems and has shown that for systems whose partial differential equations have space/time variables separable, an analytic solution is obtainable. The numerical approach used is similar to that put forward by Kranc and Sarachik (K13) for finite state systems.

10.9 TIME-OPTIMAL CONTROL

10.9.1 Problem formulation

Let Σ be a time invariant distributed system then the time optimal control problem is defined as follows.

Given

$$t_0 \in I; \qquad x(t_0), z \in X; \qquad M \text{ a real constant};$$

determine $u \in U$ such that $\|u\| \leqslant M$ almost everywhere and

(i) $x(t_1) = z$.
(ii) $t_1 = \inf\{t | x(t) = z\}$.

Let the spaces U and X be Hilbert spaces (so that the problem chosen is considerably restricted compared with the case where U, X are Banach spaces).

10.9.2 Time-optimal control: existence

Theorem 10.9. *A time-optimal control exists for the above problem.*

Proof. (Follows Balakrishnan (B3)). Let t_n be a sequence monotonically decreasing and converging to t_1. Let $u_n(\tau)$ be the corresponding elements in U so that

$$x_1 = T(t_n)x(t_0) + \int_{t_0}^{t_n} T(t_n - \tau)Bu_n(\tau)\,d\tau$$

$$= T(t_n)x(t_0) + \int_{t_0}^{t_1} T(t_n - \tau)Bu_n(\tau)\,d\tau + \int_{t_1}^{t_n} T(t_n - \tau)Bu_n(\tau)\,d\tau \quad (10.9.1)$$

Since $T(t_n)x(t_0) \to T(t_1)x(t_0)$ and the third term goes to zero, attention concentrates on the second term of equation (10.9.1).

Consider the set Ω of admissible controls (each satisfying $\|u\| \leqslant M$). Ω is closed, convex and weakly compact. Since Ω is weakly compact, there exists a weakly convergent subsequence converging to u_1 and by the closedness of Ω,

$$u_1 \in \Omega \quad \text{and hence satisfies} \quad \|u_1\| \leqslant M.$$

For every $x \in X$,

$$B^*T(t_n - \tau)^*x \in U$$

(This follows because X is a Hilbert space and $T(t)$ is a strongly continuous semi-group imply that $T(t)^*$ is a strongly continuous semi-group).

For every $x \in X$,

$$\left\langle \int_{t_0}^{t_1} T(t_n - \tau)Bu_n(\tau)\,d\tau, x \right\rangle - \left\langle \int_{t_0}^{t_1} T(t_1 - \tau)Bu_1(\tau)\,d\tau, x \right\rangle$$

$$= \int_{t_0}^{t_1} \langle u_n(\tau) - u_1(\tau), B^*T(t_1 - \tau)^*x \rangle\,d\tau$$

$$+ \int_{t_0}^{t_1} \langle u_n(\tau), B^*T(t_1 - \tau)^*(T(t_n - t_1)x - x) \rangle\,d\tau$$

By weak convergence of the sequence of u_n to u_1 the first term goes to zero. The second term has as upper bound the expression

$$k\|T(t_n - t_1)x - x\|$$

for some constant k and tends to zero as $n \to \infty$ from the strong continuity of the semi-group. Thus, for each $x \in X$,

$$\langle z, x \rangle = \left\langle T(t_1)x(t_0) + \int_{t_0}^{t_1} T(t_1 - \tau)Bu_1(\tau)\,d\tau, x \right\rangle$$

so that

$$T(t_1)x(t_0) + \int_{t_0}^{t_1} T(t_1 - \tau)Bu_1(\tau) \, d\tau = z$$

and u_1 is the optimal control.

Theorem 10.10. (A bang-bang theorem) *Assume that the set Ω is a convex neighbourhood of the origin and that $u(t)$ is a time optimal control satisfying the problem formulated above, then $u(t) \in \partial\Omega$ for almost all t in $[t_0, t_1]$.*

(*This theorem states that a time-optimal control for a distributed system is a bang-bang control.*)

Proof. See Friedman (F8).

10.9.3 Uniqueness of time-optimal control

Under an additional assumption of rotundity, uniqueness follows.

Theorem 10.11. *Let the time-optimal control problem be as stated above and let the set of admissible controls Ω be rotund, then the optimal control is unique.*

Proof. Assume u_1, u_2 are two optimal controls. Then by linearity $\frac{1}{2}(u_1 + u_2)$ is also an optimal control. From the bang-bang theorem above, u_1, u_2 and $\frac{1}{2}(u_1 + u_2)$ all belong to $\partial\Omega$ for almost all t in $[t_0, t_1]$. To say that the set is rotund is equivalent to the statement that

$$u_1, u_2, \tfrac{1}{2}(u_1 + u_2) \in \partial\Omega \Rightarrow u_1 = u_2$$

and uniqueness of the optimal control follows immediately.

10.10 OPTIMAL CONTROL OF A DISTRIBUTED SYSTEM: AN EXAMPLE

What follows is a summary of the approach taken by Hullett in reference (H12) to the modelling and control of pollution in the Delaware river.

The system state is taken to be

$$x(z; t)$$

where t = time;
z = length along the (one-dimensional) river model;
x = concentration of dissolved oxygen.

The optimal control problem is defined: Determine $u \in U$ defined on z_0, z_f; t_0, t_f such that the functional J is minimized where J is given by

$$J = \int_{t_0}^{t_f} \int_{z_0}^{z_f} Q(z)(x_d(z, t) - x(z, t))^2 \, dz \, dt + \int_{t_0}^{t_f} \int_a^b R(z)u(z, t)^2 \, dz \, dt$$

where Q, R are weighting factors and $[a, b] \subset [z_0, z_f]$.

The model equation is

$$\frac{\partial x}{\partial t} = -v\frac{\partial x}{\partial z} - k_3 x - k_1 l + k_3 x_d + u$$

$l(z, t)$ represents the biochemical oxygen demand—here assumed given—in Hullett's paper given by another partial differential equation; v represents river velocity and x_d represents the (temperature dependent) saturation value of dissolved oxygen; k_1, k_3 are constants.

$$x(z, t) \in L^2[t_0, t_f] \times L^2[z_0, z_f] = X$$

$$u(z, t) \in L^2[t_0, t_f] \times L^2[a, b] = U$$

$$[a, b] \subset [z_0, z_f]$$

Let

$$V(z, t) = -k_1 l + k_3 x_d$$

and F be the operator

$$F = -v\frac{\partial(\,)}{\partial z} - k_3(\,)$$

F is an unbounded operator so a restricted domain is defined for F by the relation

$$D(F) = \{x \in X \,|\, Fx \in X \text{ while } x \text{ satisfies boundary conditions}\}$$

An operator B is defined by the relation

$$Bu = u, \qquad z \in [a, b]$$
$$Bu = 0, \qquad \text{elsewhere}$$

so that the system model becomes

$$\frac{\partial x}{\partial t} = Fx + Bu + V \qquad\qquad (10.10.1)$$

with $x(z, 0) = x_0(z)$.

J can be written in terms of inner products

$$J = \langle x - x_d, Q(x - x_d) \rangle + \langle u, Ru \rangle$$

The optimization problem is then to minimize J subject to the constraint imposed by the model equation (10.10.1).

The semi-group $T(t)$ generated by the operator F is defined by

$$Fx = \lim_{h \to 0} \frac{T(h) - I}{h} x \qquad \text{for all } x \in D(F)$$

with F defined as above

$$T(t)x(z, t) = e^{k_3 t}x(z - vt, t)$$

The solution of the model equation (10.10.1) is

$$x(z, t) = T(t)x_0(z) + \int_{t_0}^{t_f} T(t - \tau)Bu(x, \tau)\, d\tau + \int_{t_0}^{t_f} T(t - \tau)V(z, \tau)\, d\tau \ (10.10.2)$$

Let $G : X \to X$ be defined by

$$GV = \int_{t_0}^{t_f} T(t - \tau)V(z, \tau)\, d\tau$$

while

$$P(z, t) \triangleq T(t)x_0(z) + GV$$

so that the equation (10.10.2) can be written

$$x(z, t) = GBu + P$$

Substituting this equation into the expression for J gives

$$J = \langle (GBu + P - x_d), Q(GBu + P - x_d) \rangle + \langle u, Ru \rangle$$

and, by re-arrangement,

$$J = \langle (B^*G^*QGB + R)u, u \rangle + 2\langle B^*G^*Q(P - x_d), u \rangle$$
$$+ \langle (P - x_d), Q(P - x_d) \rangle$$

The first two inner products are defined on U, the third on X.

The minimizing u for this function is given by $\hat{u}$ where

$$\hat{u} = -R^{-1}B^*G^*Q(GB\hat{u} + P - x_d)$$

or if the optimal state $\hat{x}(z, t)$ is denoted $\hat{x}$

$$\hat{u} = -R^{-1}B^*G^*Q(\hat{x} - x_d)$$

Of course, $\hat{x}$ is unknown but this problem can be overcome by making use of the adjoint equation:

$$G^*Q(\hat{x} - x_d) = \int_{t_0}^{t_f} T^*(t - \tau)Q[\hat{x} - x_d]\, d\tau$$

Put

$$w = G^*Q(\hat{x} - x_d)$$

Thus, from above,

$$\hat{u} = -R^{-1}B^*w \qquad\qquad (10.10.3)$$

to yield the following two point boundary value problem

$$\frac{\partial \hat{x}}{\partial t} = Fx - BR^{-1}B^*w + V, \qquad x(z, 0) = x_0(z)$$

$$\frac{\partial w}{\partial t} = -F^*w - Q(\hat{x} - x_d), \qquad w(x, t_f) = 0$$

where F^* is the infinitesimal generator of $T^*(t)$.

These equations, solved simultaneously, yield in conjunction with equation (10.10.3) the optimal control $\hat{u}(z, t)$.

10.11 APPROXIMATE NUMERICAL SOLUTION

For simple distributed parameter problems, it is possible, using the relations given in this chapter, to obtain analytic solutions. The calculations involve Laplace or Fourier transforms and the interested reader should start by consulting Section (4.6) of reference B7. Unfortunately, the only problems that can be solved in this way seem to be the classical problems of mathematical physics whose solutions are already well known.

More profitable from an engineering point of view is the numerical simulation of distributed parameter problems using the methods of this chapter as a framework. This approach is described briefly below. A detailed derivation including an identification procedure and giving numerical results is in reference (L5).

Let the state space X and the control space U be Hilbert spaces. A very general model can be postulated as $x(z) = E\,u(t)$.

Let $\{\gamma_i\}$ be an orthonormal basis for X and $\{\xi_i\}$ be an orthonormal basis for U.

Define

$$u_j = \langle u, \xi_j \rangle \quad \text{so that} \quad u = \sum_j u_j \xi_j$$

and define similarly $x_i = \langle x, \gamma_i \rangle$.

Then

$$x(z) = E \sum_j u_j \xi_j = \sum_j u_j E\xi_j$$

Taking inner products with γ_i,

$$\langle x, \gamma_i \rangle = \langle \sum_j u_j E\xi_j, \gamma_i \rangle$$

which can be written

$$x_i = \sum_j u_j \langle E\xi_j, \gamma_i \rangle$$

or

$$x = \sum_i x_i \gamma_i = \sum_i \sum_j u_j \langle E\xi_j, \gamma_i \rangle \gamma_i$$

Denote the scalars $\langle E\xi_j, \gamma_i \rangle$ by c_{ji} then we have a model

$$x = \sum_i \sum_j u_j c_{ji} \gamma_i$$

In practice, the summations are truncated, so that the system is modelled by a finite matrix.

10.12 EXERCISES

(1) Let $\{T_i\}$ be a sequence of operators on a normed linear space X. Show by counterexample that even though

$$\{T_i x\} \to Tx \qquad \text{for all } x \in X$$

it does not follow that

$$\{T_i\} \to T$$

(2) Let

$$X = C[0, \infty), \qquad x(t) \in X$$

Show that the differential operator A defined by $Ax(t) = \mathrm{d}x(t)/\mathrm{d}t$ is the infinitesimal generator of $T(h)$ where

$$T(h)(x(t)) = x(t + h), \qquad h \geqslant 0$$

Show that $\{T(h)\}$ is a strongly continuous semi-group.

(3) Let a distributed parameter system be represented by the equation $\dot{x} = Ax, x \in X$ where X is a Banach space, $x(0)$ given. Let A be the infinitesimal generator of a strongly continuous semi-group $\{T(t), t \geqslant 0\}$, $T: X \to X$.

Show that:

(i) $x(0) \in D(A) \Rightarrow T(t)x(0) \in D(A), \forall t \geqslant 0$.

(ii) $T(t)x(0)$ is a solution of the given equation.
(iii) $D(A)$ is dense in X.
(iv) A is a closed operator.

Is A necessarily bounded?

(4) Consider the equation $\dot{x} = Ax$, $x(0)$ given. Suppose that X is the set of real functions continuous on a finite interval $[a, b]$.

Let A be the operator $A : X \to X$ given by

$$A = \frac{\partial(\ )}{\partial z}$$

i.e. the domain of A is the subset of X consisting of functions whose first derivative is continuous.

Is $D(A)$ dense in X? Is A a closed operator?

(5) Investigate the following equations. Put each in the form required for the methods of this chapter to be applicable. Define all the spaces and operators. Characterize the solutions in terms of an operator $T(t)$.

(a) $\dfrac{\partial x(t, z)}{\partial t} = \dfrac{\partial x(t, z)}{\partial z}$, $x(0, z)$ given.

(b) The heat diffusion equation

$$\frac{\partial x(t, z)}{\partial t} = \frac{\partial^2 x(t, z)}{\partial z^2}, \qquad x(0, z), \ \frac{\partial x}{\partial z}(0, z) \text{ given.}$$

(c) The wave equation

$$\frac{\partial^2 x(t, z)}{\partial t^2} = \frac{\partial^2 x(t, z)}{\partial z^2}, \qquad x(0, z), \ \frac{\partial x}{\partial z}(0, z) \text{ given.}$$

(d) The diffusion equation with forcing

$$\frac{\partial x(t, z)}{\partial t} = \frac{\partial^2 x(t, z)}{\partial z^2} + u(t, z)$$

$$x(0, z), \frac{\partial x}{\partial z}(0, z) \text{ given,}$$

$$u(t, z) \text{ defined for all } t \text{ and all } z.$$

Glossary of Symbols

Some symbols can take on meanings different from those given in the list—such meanings are defined locally in the text.

$\mathscr{A}$	The attainable set
$\mathscr{C}$	The set of convergent sequences
$\mathscr{C}_0$	The set of sequences convergent to zero
$\mathscr{I}(z)$	The imaginary part of z
$\mathscr{L}(z)$	The Laplace transform of z
$\mathscr{R}$	The reachable set
$\mathscr{R}(z)$	The real part of z
$\mathbb{R}^n$	Real n dimensional space
A, B, C	System matrices or system operators
$B(X, Y)$	The family of bounded linear operators from X to Y
$B(X)$	The family of bounded linear operators from X to X
$BV[a, b]$	The set of functions of bounded variations defined on $[a, b]$
$C[a, b]$	The set of functions continuous on $[a, b]$
$D(A)$	The domain of A
H	A hyperplane
I	The identity operator *or* used in general to represent a domain
J	The cost index for an optimization problem
K	A convex set *or* the set of real scalars
$L(x_1, \ldots, x_n)$	The linear subspace generated by $x_1, \ldots, x_n$
$R(s, A)$	The resolvent of the operator A
U	The space of control functions
V	The dual space for U
X	The state space
Y	The output space
Z	The set of integers
Z^+	The set of positive integers
D_{ij}	Defines a matrix D by giving the value of each element

G_A	The graph of A
N_A	The null space of the mapping A
$N_\varepsilon(x)$	An epsilon neighbourhood of x
A^T	The transpose of matrix A
A^*	The complex conjugate of the matrix A *or* the adjoint operator of the operator A
E^0	The interior of the set E
$\bar{E}$	The closure of the set E
$E^\perp$	The set of elements orthogonal to the set E
E'	The complement of the set E
L^p	The space of (equivalence classes of) functions whose pth power is Lebesgue integrable
$(L^p)^r$	An r valued L^p space
$X \times Y$	The product set of X and Y
$d(\ ,\ )$	A metric
e^A or $\exp(A)$	Used interchangeably
k cube	$= \left\{ u \middle\| u = \begin{pmatrix} u_1 \\ \vdots \\ u_r \end{pmatrix}, |u_i| \leqslant k, i = 1, \ldots, r \right\}$
l^p	The space of sequences whose pth power is summable
p, q	Suffices such that $p + q = pq$
s	The complex variable associated with Laplace transformation
x, u, v, y	Mostly used to represent vectors
Γ	A transformation semi-group
Λ	The diagonal matrix of eigenvalues
Σ	The notation for an abstract system
Φ	The transition matrix
Ψ	The forced solution matrix (see Section 6.3)
Ω	The set of admissible controls
$\beta(X)$	The closed unit ball in X
$\partial(E)$	The boundary of the set E
η	The output mapping
λ	An eigenvalue
$\mu(E)$	The measure of the set E
$\rho(A)$	The resolvent set of A
$\sigma(A)$	The spectrum of A
τ	A topology (also used as a time variable)
ϕ	The transition mapping
χ	The characteristic function
ψ	A row of the matrix Ψ
$\varnothing$	The empty set
$\{\ \}$	A set or sequence

$(\ ,\ )$	Open interval
$[\ ,\ ]$	Closed interval
$[\ :\ ]$	Line segment
$\langle\ .\ \rangle$	Inner product
$(\ )_{ij}$	An individual element of a matrix
$\oplus$	Direct sum
$\|\cdot\|$	Norm
$|\cdot|$	Absolute value
$x \perp y$	x is orthogonal to y
$\det(A)$	The determinant of A
$\dim(X)$	The dimension of the space X
$\mathrm{grad}(x)$	The gradient of the vector x
inf	Infimum
sign (q)	Defined by sign $(q) = 0$ if $q = 0$
	$\qquad\qquad\qquad\ = 1$ if $q > 0$
	$\qquad\qquad\qquad\ = -1$ if $q < 0$
sup	Supremum
$\Rightarrow$	Implies
$\Leftarrow$	Is implied by
$\Leftrightarrow$	Implies and is implied by
$\overline{\langle\ ,\ \rangle}$	The complex conjugate of $\langle\ ,\ \rangle$
$\triangleq$	Equal to by definition
$\tilde{f}$	Extremum
$\dot{q}$	The time derivative of q
ε, δ	are used in the conventional sense to represent small quantities
Σ, Π	represent summation and continued multiplication respectively

References and Further Reading

General mathematics references

Here are listed, outside the main body of references, a number of books that are recommended for consolidation of the reader's mathematical background. There are very many books available in these areas and this selection is based on personal preference.

Gemignani, M.C., "Elementary Topology". Addison-Wesley, Reading, Mass., 1967.
  Uses many diagrams to support the text.
Hafstrom, J.E., "Introduction to Analysis and Abstract Algebra". W.B. Saunders Co., Philadelphia and London, 1967.
  A readable, detailed introduction to many of the basic concepts of sets and functions.
Jameson, G.J.O., "Topology and Normed Spaces". Chapman and Hall, London, 1974.
  An introductory text on normed spaces.
Kolmogorov, A.N. and Fomin, S.V., "Elements of the Theory of Functions and Functional Analysis". Graylock Press, Rochester N.Y., 1957.
  I. Metric and normed spaces.
  II. Measure, Lebesgue integral, Hilbert space.
Munroe, M.E., "Introduction to Measure and Integration". Addison-Wesley, Reading, Mass., 1953.
Protter, M.H. and Morrey, C.B., "A First Course in Real Analysis". Undergraduate texts in mathematics. Springer-Verlag, New York, 1977.
Rudin, W., 'Principles of Mathematical Analysis". McGraw-Hill, New York, 1964.

Books on control theory

Takahashi (T1), Ogata (O1), Padulo (P1), and Zadeh (Z1) are (in increasing order of abstractness) comprehensive texts on control theory. Polak (P5) and Desoer (D10) cover linear control theory very concisely.

Books on functional analysis

Day (D7), Holmes (H10) cover normed spaces.

Akhiezer (A2), Berberian (B9), Dunford (D12), Hille (H9), Riesz (R2), Yosida (Y1) cover functional analysis and operator theory with D12 being a reference treatise on operator theory and H9 being the same for semi-group theory.

Balakrishnan (B7), Caianiello (C2), Curtain (C7), Hermes (H8), Lions (L9), Luenberger (L11), Porter (P9), cover applications of functional analysis to control or to related areas.

REFERENCES

The concepts described in this book were developed over a long period by many mathematicians and control theorists. Papers are included in the following list, in addition to those specifically cited, to allow this historical development to be appreciated and to give further in-depth support to the text.

A1   AHMED, N.U. and TEO, K.L., Necessary conditions for optimality of Cauchy problems for parabolic partial differential systems. *SIAM J. Control* **13**, No. 5, 981–993 (1975).

A2   AKHIEZER, N.J. and GLAZMAN, T.N., "Theory of Linear Operators in Hilbert Space", Vols. I and II. Frederick Ungar, New York, 1961.

A3   AKHIEZER, N.J. and KREIN, M. (Eds.), "Some Questions in the Theory of Moments". Translation of a 1932 Russian work. American Mathematical Society, Providence, Rhode Island, 1962.

A4   ALVARADO, F.L. and MUKUNDAN, R., An optimization problem in distributed parameter systems. *Internat. J. Control* **9**, No. 6, 665–677 (1969).

A5   ANTOSIEWICZ, H.A., Linear control systems. *Arch. Rational. Mech. Anal.,* **12**, 313–324 (1963).

A6   AOKI, M., Minimum norm problems and some other control system optimization techniques. *In* "Advances in Control Systems" (Ed. C.T. Leondes), Vol. 2. Academic Press, New York, 1965.

A7   AXELBAND, E.I., Function space methods for the optimum control of a class of distributed parameter systems. Preprints Joint Automat. Control Conf. of the American Control Council, Troy, New York, pp. 374–380. Published IEEE, 1965.

B1   BALAKRISHNAN, A.V., Linear systems with infinite dimensional state spaces. Proc. Symp. Systems Theory, pp. 69–98. Polytechnic Press of the Polytechnic Institute of Brooklyn, New York, 1965.

B2   BALAKRISHNAN, A.V., Semi-group theory and control theory. Proc. Internat. Federation Information Processing, pp. 157–163. Spartan Books, Washington D.C., 1965.

B3   BALAKRISHNAN, A.V., Optimal control problems in Banach spaces. *SIAM J. Control* **3**, No. 1, 152–180 (1965).

B4   BALAKRISHNAN, A.V., On the state space theory of linear systems. *J. Math. Anal. Appl.* **14**, 371–91 (1966).

B5   BALAKRISHNAN, A.V., Stochastic control—a function space approach. National Science Foundation Regional Conference on Control Theory, University of Maryland, August 1971. Reprinted *SIAM J. Control* **10**, No. 2, 285–297 (1972).

147

B6 BALAKRISHNAN, A.V., "Introduction to Optimization Theory in a Hilbert Space". Springer Lecture Notes in Operations Research and Mathematics, Vol. 42. Springer-Verlag, Berlin, 1971.

B7 BALAKRISHNAN, A.V., "Applied functional analysis". Springer-Verlag, New York, 1976.

B8 BARAS, J.S., BROCKETT, R.W. and FUHRMANN, P.A., State space models for infinite-dimensional systems. *IEEE Trans. Automatic Control* AC-19, No. 6, 693–700 (1974).

B9 BERBERIAN, S.K., "Lectures in Functional Analysis and Operator Theory". Springer-Verlag, New York, 1974.

B10 BISHOP, E. and PHELPS, R.R., The support functionals of a convex set. *Proc. Amer. Math. Soc.* 2, 27–39 (1963).

B11 BROCKETT, R.W., "Finite Dimensional Linear Systems". Wiley, New York, 1970.

B12 BROGAN, W.L., Theory and application of optimal control for distributed parameter systems. *Automatica—J. IFAC* 4, No. 3 107–137 (1967).

B13 BROGAN, W.L., Optimal control theory applied to systems described by partial differential equations. *In* "Advances in Control Systems" (Ed. C.T. (Leondes), Vol. 6. Academic Press, New York and London, 1968.

B14 BUTKOVSKY, A.G., The method of moments in the theory of optimal control of systems with distributed parameters. *Automat. Remote Control* 24, No. 9, 1106–1113 (1963).

B15 BUTKOVSKY, A.G., Optimal control of systems with distributed parameters. Proc. 2nd Congress IFAC, Basle, Switzerland (1963), Vol. 2, pp. 333–338. Butterworths, London, and Oldenbourg, Munich, 1964.

B16 BUTKOVSKY, A.G., "Theory of Optimal Control for Distributed Parameter Systems. "Nayka" Publishers, Moscow, 1965.

B17 BUTKOVSKY, A.G. and LERNER, A.Y., Optimal control of distributed parameter systems. *SIAM J. Control* 6, No. 3, 437–476 (1968).

B18 BUTKOVSKY, A.G., "Distributed Control Systems". American Elsevier, New York, 1969.

C1 CADSOW, J.A., A study of minimum norm control for sampled data systems. Preprints Joint Automat. Control Conf. of the American Control Council, Troy, New York. Published IEEE, Paper XIX-2, 1965.

C2 CAIANIELLO, E.R. (Ed.), "Functional Analysis and Optimization". Academic Press, New York and London, 1966.

C3 CESARI, L., Existence theorems for multi-dimensional problems of optimal control. University of Michigan report, March 1966.

C4 CESARI, L., Semi-normality and upper semi-continuity in optimal control. *J. Optim. Theory Appl.* 6, No. 2, 114–137 (1970).

C5 CESARI, L., Existence theorems for abstract multi-dimensional control problems. *J. Optim. Theory Appl.* 6, No. 3, 210–236 (1970).

C6 CLARKSON, J.A., Uniformly convex spaces. *Trans. Amer. Math. Soc.,* 4, 394–414 (1936).

C7 CURTAIN, R.F. and PRITCHARD, A.J., "Functional Analysis in Modern Applied Mathematics". Mathematics in Science and Engineering, Vol. 132 Academic Press, London and New York, 1977.

D1 DATKO, R. and ANVARI, M., The existence of optimal controls for a performance index with a positive integrand *in* "Differential Equations and Dynamical Systems" (Ed. J.K. Hale and J.P. La Salle). Academic Press, New York and London, 1967.

D2 DATKO, R., A linear control problem in an abstract Hilbert space. *J. Differential equations* **9**, 346–359 (1971).

D3 DAY, M.M., The space l^p with $0 < p < 1$. *Bull. Amer. Math. Soc.* **46**, 816–23 (1940).

D4 DAY, M.M., Reflexive Banach spaces not isomorphic to uniformly convex spaces. *Bull. Amer. Math. Soc.* **47**, 313–317 (1941).

D5 DAY, M.M., Some more uniformly convex spaces. *Bull. Amer. Math. Soc.* **47**, 504–507 (1941).

D6 DAY, M.M., Strict convexity and smoothness of normed spaces. *Trans. Amer. Math. Soc.* **48**, 516–528 (1955).

D7 DAY, M.M., "Normed Linear Spaces". Academic Press, New York and London , 1962.

D8 DELFOUR, M.C. and MITTER, S.K., Controllability, observability and optimal feedback control of affine hereditary differential systems. *SIAM J. Control* **10**, No. 2, 298–328 (1972).

D9 DEM'YANOV, V.F. and RUBINOV, A.M., The minimization of a smooth convex functional on a convex set. *SIAM J. Control* **5**, No. 2, 280–294 (1967).

D10 DESOER, C.A., "Notes for a Second Course in Linear Systems". Van Nostrand Reinhold, New York, 1970.

D11 DIESTEL, J., "Geometry of Banach spaces". Lecture Notes in Mathematics, Vol. 485. Springer-Verlag, Berlin, 1975.

D12 DUNFORD, N. and SCHWARTZ, J.T., "Linear operators". Vol. I (General), Wiley Interscience, New York, 1958. Vol. II (In Hilbert Space), Wiley Interscience, 1963.

D13 DVORETSKY, A., Some results on convex bodies and Banach spaces. Proc. Internat. Symp. Linear Spaces, Jerusalem, 1960.

D14 DYER, D.A.J. and HILLER, J., Invariance, adjoints and the necessary conditions for an extremum. *Internat. J. Control* **16**, No. 6, 1021–1027 (1972).

E1 ERZBERGER, H. and KIM, M., Optimum boundary control of distributed parameter systems. *Inform. and Control* **9**, 265–278 (1966).

E2 ERZBERGER, H. and KIM, M., Optimum distributed parameter systems with distributed control. *Proc. IEEE* **54**, 714–715 (1966).

F1 FALB, P.L., Infinite dimensional control problems: on the closure of the set of attainable states for linear systems. *J. Math. Anal. Appl.* **9**, 12–22 (1964).

F2 FELLER, W., The parabolic differential equation and the associated semi-group of transformations. *Ann. of Math.* **55**, No. 2, 468–519 (1952).

F3 FILLIPOV, A.F., On certain questions in the theory of optimal control. *SIAM J. Control* **1**, No. 1, 76–84 (1962).

F4 FISHER, S.W. and JEROME, J.W., "Minimum Norm Extremals in Function Spaces". Lecture Notes in Mathematics, Vol. 479. Springer-Verlag, Berlin, 1975.

F5 FLEMING, W.H. and NISIO, M., On the existence of optimal stochastic controls. *J. Math. Mech.* **15**, 777–794 (1966).

F6 FREEMAN, E.A., Stability of linear constant multi-variable systems. *Proc. IEE* **120**, No. 3 (1973).

F7 FREEMAN, E.A., Some control system stability and optimality results obtained via functional analysis. *In* "Recent Mathematical Developments in Control" (Ed. D.J. Bell). Academic Press, London and New York, 1973.

F8 FRIEDMAN, A., Optimal control in Banach spaces. *J. Math. Anal. Appl.* **18**, 35–55 (1967).

G1 GABASOV, R. and KIRRILLOVA, F.M., Optimisation of linear systems by the methods of functional analysis. *J. Optim. Theory Appl.* **8**, No. 2, 77–99 (1971).

G2   GILBERT, E.G., Controllability and observability in multivariable control systems. *SIAM J. Control* **1**, No. 2, 128–151 (1963).

G3   GILBERT, E.G., The Neustadt Algorithms and other convexity methods for the computation of optimal control. Preprints Joint Automat. Control Conf. of the American Control Council, Ohio State University, 1973. Publ. IEEE, paper 1–4, 1973.

G4   GIRSANOV, I.V., "Lectures on the Mathematical Theory of Extremum Problems". Springer Lecture Notes in Economics and Mathematical Systems, Vol. 67. Springer-Verlag, Berlin, 1972.

G5   GOLDSTEIN, A.A., Minimising functionals on normed linear spaces. *SIAM. J. Control* **4**, No. 1, 81–89 (1966).

G6   GOODSON, R.E. and KHATRIC, H.C., Optimal control of systems with distributed parameters. *Trans. ASME Ser. D. J. Basic Eng.* **88**, No. 2, 337–342 (1966).

H1   HALKIN, H., Topological aspects of optimal control. *Contr. Diff. Eqns.* **3**, 377–385 (1964).

H2   HALKIN, H. and NEUSTADT, L.W., General necessary conditions for optimization problems. *Proc. Nat. Acad. Sci. USA* **56**, 1066–1071 (1966).

H3   HALKIN, H., Mathematical foundations of systems optimization. *In* "Topics in Optimization" (Ed. G. Leitmann), Academic Press, New York and London, 1967.

H4   HALKIN, H., Lucien Neustadt's contributions to the theory of necessary conditions for optimization problems. Preprints Joint Auto. Control Conf. of the American Automatic Control Council, Ohio State University, 1973. Publ. IEEE, paper 1–2, 1973.

H5   HALMOS, P., The range of a vector measure. *Bull. Amer. Math. Soc.* **54**, 416–421 (1948).

H6   HEMMING, F. and VANDELINDE, V.D., An optimal control problem in Banach space. *J. Math. Anal. Appl.* **39**, No. 3, 647–654 (1972).

H7   HERMES, H., On the closure and convexity of attainable sets in finite and infinite dimensions. *SIAM J. Control* **5**, No. 3, 409–417 (1967).

H8   HERMES, H. and LA SALLE, J.P., Functional analysis and time-optimal control. Academic Press, New York and London, 1969.

H9   HILLE, E. and PHILLIPS, R., "Functional Analysis and Semi-groups". Publication, Vol. 31. Providence, Rhode Island, 1957. American Mathematical Society.

H10   HOLMES, R.B., "Geometric Functional Analysis and its Applications". Springer-Verlag, New York, 1975.

H11   HSIEH, H.C. and NESBITT, R.A., Functional analysis and its applications to mean-square error problems. *In* "Advances in Control Systems" (Ed. C.T. Leondes) Vol. 14. Academic Press, New York, 1966.

H12   HULLETT, W., Optimal estuary aeration: an application of distributed parameter control theory. *Appl. Math. Optim.* **1**, No. 1, 20–63 (1974).

I1   *International Atomic Energy Agency*, Control theory and topics in functional analysis. Vol. I, II, III. (Lectures presented at an international seminar in Trieste 11th September to 24th November 1974). Publ. Internat. Atomic Energy Agency, Vienna, 1976.

J1   JACOBS, M.Q., Linear optimal control problems. *SIAM J. Control* **5**, No. 3. 418–437 (1967).

J2   JAMES, R.C., Reflexivity and the supremum of linear functionals. *Ann. of Math.* **66**, No. 2, 159–169 (1957).

J3   JAMES, R.C., Characterisations of reflexivity. *Studia Math.* **23**, 205–216 (1964).

J4   JOHNSSON, L., Distributed parameter systems. An annotated bibliography to April 1971. U. of California, Los Angeles (UCLA Eng. 7143), 1972.

J5   JURDJEVIC, V., Abstract control systems—controllability and observability. *SIAM J. Control* **8**, No. 3, 424–439 (1970).

K1   KALMAN, R.E., On the general theory of control systems. Proc. 1st Congress IFAC, Moscow, USSR, 1960, Vol. 1, pp. 481–492. Butterworths, London, 1961.

K2   KALMAN, R.E., Canonical structure of linear dynamical systems. *Proc. Nat. Acad. Sci. USA* **48**, 596–600 (1962).

K3   KALMAN, R.E., Mathematical description of dynamical systems. *SIAM J. Control* **1**, No. 2, 152–192 (1963).

K4   KALMAN, R.E., HO, Y.C. and NARENDRA, K.S., "Controllability of Linear Dynamical Systems". Wiley, New York, 1963.

K5   KALMAN, R.E., Contributions to linear system theory. *Internat. J. Engrg. Sci.* **3**, 141–171 (1965).

K6   KALMAN, R.E., On the structural properties of linear constant multi-variable systems. Proc. 3rd Congress IFAC, 1966. Publ. Inst. Mechanical Engnrs. (U.K.), Paper 6A, 1967.

K7   KIM, M. and ERZBERGER, H., On the design of optimum distributed parameter systems with boundary control function. *IEEE Trans. Automatic Control* **AC-12**, 22–37 (1967).

K8   KIM, M. and GAJWANI, S.H., A variational approach to optimum distributed parameter systems. *IEEE Trans. Automatic Control* **AC-13**, 191–194 (1968).

K9   KIRILLOVA, F.M., Applications of functional analysis to the theory of optimal processes. *SIAM J. Control* **5**, No. 1, 25–49 (1967).

K10   KISYNSKI, J., Semi-groups of operators and some of their applications to partial differential equations. Lecture presented at the international seminar course, Trieste, 1974, and published in reference I1, volume III.

K11   KLEE, V., Separation and support properties of convex sets—a survey. *In* "Control Theory and the Calculus of Variations" (Ed. A.V. Balakrishnan). Academic Press, New York and London, 1969.

K12   KNOWLES, G., Lyapunov vector measures. *SIAM J. Control* **13**, No. 2, 294–303 (1975).

K13   KRANC, G.M. and SARACHIK, P.E., An application of functional analysis to the optimal control problem. *Trans. ASME Ser. D. J. Basic Eng.* **85**, 143–150 (1963).

K14   KRASOWSKII, N.N., On the theory of optimum regulation. *Automat. Remote Control* **18**, 1005–1016 (1959).

K15   KRASOWSKII, N.N., On the theory of optimum control. *J. Appl. Math. Mech.* **23**, 899–919 (1959).

K16   KREIN, M., The L problem in abstract linear normed spaces. Published in reference A3.

K17   KREINDLER, E., Contributions to the theory of time-optimal control. *J. Franklin Inst.* **275**, 314–344 (1963).

K18   KREINDLER, E. and SARACHIK, P.E., On the concepts of controllability and observability of linear systems. *IEEE Trans. Automatic Control* **AC-9**, 129–136 (1964).

L1   LADAS, G.S. and LAKSHMIKANTHAM, V., "Differential equations in abstract spaces". Academic Press, New York and London, 1972.

L2   LA SALLE, J.P., Study of the basic principle underlying the "bang-bang" servo. Goodyear Aircraft Corp. Report GER-5518, 1953.

L3   LA SALLE, J.P., Time optimal control systems. *Proc. Nat. Acad. Sci. USA* **45**, 573–577 (1959).

L4   LA SALLE, J.P., The time optimum control problem. *In* "Contributions to the Theory of Nonlinear Oscillations" (Ed. S. Lefschetz), Vol. 5. Princeton University Press, Princeton N.J., 1960.

L5   LAU, C.C. and LEIGH, J.R., Modelling of an industrial distributed parameter process. Proc. 5th IFAC Symposium on Identification and System Parameter Estimation. Darmstadt, Federal Republic of Germany, 24–28th September 1979. Publ. Pergamon Press, paper A4.4, 1980.

L6   LEE, E.B., Mathematical aspects of the synthesis of linear minimum response time controllers. *IEEE Trans. Automat. Control* **AC-5**, 283–289 (1960).

L7   LEE, E.B., Geometric properties of optimal controllers for linear systems. *IEEE Trans. Automat. Control* **AC-8**, 379–381 (1963).

L8   LINDENSTRAUSS, J. and PHELPS, R., Extreme point properties of convex bodies in reflexive Banach spaces. *Israel J. Math.* **6**, 39–48 (1968).

L9   LIONS, J.L., Control problems and partial differential equations. *Proc. Conf. on Math. Theory of Control*, University of South California, Los Angeles, 1967.

L10 LIONS, J.L., "Optimal Control of Systems Governed by Partial differential equations". Springer-Verlag, Berlin, 171.

L11 LUENBERGER, D.G., "Optimization by Vector Space Methods". Wiley, New York, 1969.

M1   MARKUS, L. and LEE, E.B., On the existence of optimal controls. *Trans. ASME Ser. D. J. Basic Eng.* **84**, 13–22 (1962).

M2   MARZOLLO, A., Controllability and optimization. Lectures held at the International Centre for Mechanical Sciences, Udine. Springer-Verlag, New York, 1972.

M3   MESAROVIC, M.D., "General Systems Theory—Mathematical Foundations". Academic Press, New York and London, 1975.

M4   MIRANKER, W.L., Approximate controllability for distributed linear systems. *J. Math. Anal. Appl.* **10**, 378–387 (1965).

M5   MIZUKAMI, K., Applications of functional analysis to optimal control problems. Lecture presented at the International Seminar Course, Trieste, 1974, and published in reference I1, volume II.

N1   NEUSTADT, L.W., Synthesising time optimal control systems. *J. Math. Anal. Appl.* **1**, 484–492 (1960).

N2   NEUSTADT, L.W., Minimum effort control systems. *SIAM J. Control* **1**, 16–31 (1962).

N3   NEUSTADT, L.W. and PAIEWONSKY, B.H., On synthesising optimal controls. Proc. 2nd Congress IFAC, Basle, 1963. pp. 283–291. Publ. Butterworths, London and Oldenbourg, Munich, 1964.

N4   NEUSTADT, L.W., The existence of optimal controls in the absence of convexity conditions. *J. Math. Anal. Appl.* **7**, 110–117 (1963).

N5   NEUSTADT, L.W., Optimal control problems as extremal problems in a Banach space. Proc. Symp. Systems Theory, pp. 205–214. Polytechnic Press of the Polytechnic Institute of Brooklyn, New York, 1965.

N6   NEUSTADT, L.W., An abstract variational theory with application to a broad class of optimization problems: Pt. 1. General Theory, Pt. 2, Applications. *SIAM J. Control* **4**, No. 4, 505–527 (1966); **5**, No. 1, 90–137 (1967).

N7   NISHIURA, T., On an existence theorem for optimal control. *SIAM J. Control* **5**, No. 4, 532–544 (1967).

O1   OGATA, K., "State Space Analysis of Control Systems". Prentice-Hall, Englewood-Cliffs, N.J., 1967.

O2   OKAZAWA, N., A perturbation theorem for linear contraction semigroups on reflexive Banach spaces. *Proc. Japan Acad. Sci.* **47**, 947–949 (1971).

O3   OLECH, C., Existence theorems for optimal problems with vector valued cost functions. *Trans. Amer. Math Soc.* **36**, 159–180 ⟨1969⟩.

P1   PADULO, L. and ARBIB, M.A., "System Theory". W.B. Saunders, Philadelphia, 1974.

P2   PETTIS, B.J., A proof that every uniformly convex space is reflexive. *Duke Math. J.* **5**, 249–523 (1939).

P3   PHELPS, R.R., Uniqueness of Hahn-Banach extensions and unique best approximation. *Trans. Amer. Math. Soc.* **95**, 238–255 (1960).

P4   POLAK, E. and DEPARIS, M., An alogrithm for minimum energy control. *IEEE Trans. Automatic Control* **AC-14**, 367–377 (1969).

P5   POLAK, E. and WONG, E., "Notes for a First Course in Linear Systems". Van Nostrand Reinhold, New York, 1970.

P6   PONTRYAGIN, L.S., BOLTYANSKII, V.G., GAMKRELIDZE, R.V. and MISHCHENKO, E.F., "The Mathematical Theory of Optimal Processes". Pergamon Press, Oxford, 1964.

P7   PORTER, W.A., A new approach to a general minimum energy problem. Preprints Joint Automat. Control Conf. of the American Automatic Control Council, Hartford, California. pp. 228–232. Publ. IEEE, 1964.

P8   PORTER, W.A. and WILLIAMS, J.P., Minimum effort control of linear dynamic systems. Tech. Report 5892-20-F, Contract AF 33 (657)-11501, 1964.

P9   PORTER, W.A., "Modern Foundations of System Engineering". Macmillan New York, 1966.

P10 PORTER, W.A. and WILLIAMS, J.P., A note on the minimum effort control problem. *J. Math. Anal. Appl.* **13**, 251–264 (1966).

P11 PORTER, W.A. and WILLIAMS, J.P., Extensions of the minimum effort control problem. *J. Math. Anal. Appl.* **13**, 536–549 (1966).

P12 PORTER, W.A., On the optimal control of distributive systems. *SIAM J. Control* **4**, No. 3, 466–472 (1966).

P13 PORTER, W.A., A basic optimization problem in linear systems. *Mathematical Systems Theory* **5**, No. 1, 20–44 (1971).

P14 PORTER, W.A. and ZAHM, C.L., Basic concepts in system theory. Report number 011321-1-T. Systems Engineering Laboratory, University of Michigan, 1972.

P15 PRADO, G., Observability, estimation and control of distributed parameter systems. MIT report (ESL-R-457), 1971.

P16 PSHENICHNIY, B.N., Linear optimal control problems. *SIAM J. Control* **4**, No. 4, 577–593 (1966).

P17 PSHENICHNIY, B.N., "Necessary Conditions for an Extremum". Marcel Dekker, New York, 1971.

R1   REID, W.T., Ordinary linear differential operators of minimum norm. *Duke Math. J.* **29**, 591–606 (1962).

R2   RIESZ, F. and SZ-NAGY, B., "Functional Analysis". Frederick Ungar, New York, 1955.

R3   ROBINSON, A.C., A survey of optimal control of distributed parameter systems. Aerospace Research Laboratory report (ARL 69-0177) (AD 701-732), 1969.

R4 ROXIN, E., On the existence of optimal controls. *Michigan Math. J.* **9**, 109–119 (1962).

R5 RUBIO, J. E., "The Theory of Linear Systems". Academic Press, New York and London, 1971.

S1 SAKAWA, Y., Solution of an optimal control problem in a distributed parameter system. *IEEE Trans. Automat. Control* **AC-9**, 420–426 (1964).

S2 SARACHIK, P.E., Functional analysis in automatic control. *IEEE Internat. Convention Record* **6**, 96–107 (1965).

S3 SCHMAEDEKE, W.W., The existence theory of optimal control systems. *In* "Advances in Control Systems" (Ed. C.T. Leondes), Vol. 3. Academic Press, New York and London, 1966.

S4 SILVERMAN, L.M. and ANDERSON, B.D.O., Controllability, observability and stability of linear systems. *SIAM J. Control* **6**, No. 1, 121–130 (1968).

S5 SLEMROD, M., A note on complete controllability and stabilisability for linear control systems in Hilbert spaces. *SIAM J. Control* **12**, No. 3, 500–508 (1974).

S6 STODDART, A.W., Existence of optimal controls. *Pacific J. Math.* **20**, No. 1, 167–177 (1967).

S7 STOER, J. and WITZGALL, C., "Convexity and Optimization in Finite Dimensions". Springer-Verlag, Berlin, 1970.

S8 STOREY, C., Stability and controllability. *In* "Recent Mathematical Developments in Control" (Ed. D.J. Bell). Academic Press, London and New York, 1973.

S9 SUNDARESAN, K., Smoothness classification of reflexive spaces. *In* "The Geometry of Metric and Linear Spaces" (Ed. L.M. Kelly). Springer-Verlag, 1975.

S10 SWIGER, J.M., Application of the theory of minimum-normed operators to optimum control systems problems. *In* "Advances in Control Systems" (Ed. C.T. Leondes), Vol. 3. Academic Press, New York, 1966.

T1 TAKAHASHI, Y., RABINS, M.J. and AUSLANDER, D.M., "Control and Dynamic Systems". Addison-Wesley, Reading, Mass, 1970.

T2 TRIGGIANI, R., Controllability and observability in Banach space with bounded operators. *SIAM J. Control* **13**, No. 2, 462–492 (1975).

V1 VAYSBORD, E.M., Optimal control of systems with distributed parameters. *Eng. Cybernetics* **4**, No. 5, 163–167 (1966).

V2 VINTER, R.B., A generalisation to dual Banach spaces of a theorem by Balakrishnan. *SIAM J. Control* **12**, No. 1, 150–166 (1974).

W1 WANG, P.K.C. and TUNG, F., Optimum control of distributed parameter systems. *Trans. ASME Ser. D. J. Basic Eng.* **86**, No. 1, 66–79 (1964).

W2 WANG, P.K.C., Control of distributed parameter systems. *In* "Advances in Control Systems" (Ed. C.T. Leondes), Vol. 1. Academic Press, New York and London, 1964.

W3 WANG, P.K.C., On the feedback control of distributed parameter systems. *Internat. J. Control* **3**, No. 3, 255–273 (1966).

W4 WANG, P.K.C., Asymptotic stability of distributed parameter systems with feedback controls. *IEEE Trans. Automatic Control* **AC-11**, No. 1, 46–54 (1966).

W5 WANG, P.K.C., Control of a distributed parameter system with a free boundary. *Internat. J. Control* **5**, No. 4, 317–329 (1967).

W6 WANG, P.K.C., Theory of stability and control for distributed parameter systems. *Internat. J. Control* **7**, No. 2, 101–116 (1969).

W7 WARGA, J., "Optimal Control of Differential and Functional Equations". Academic Press, New York and London, 1972.

W8   WEISS, L. and KALMAN, R.E., Contributions to linear system theory. *Internat. J. Engrg. Sci.* **3**, 141–171 (1965).

W9   WIBERG, D.M., Feedback control of linear distributed systems. *Trans. ASME Ser. D. J. Basic Eng.* **89**, 379–384 (1967).

W10 WILLEMS, J.C., "The Analysis of Feedback Systems". MIT Press, Cambridge, Mass, 1971.

W11 WILLEMS, J.C., "Stability Theory of Dynamical Systems". Nelson, London, 1970.

W12 WINDEKNECHT, J.G., "General Dynamical Processes". Academic Press, New York and London, 1971.

W13 WOLOVICH, W.A., "Linear Multivariable Systems". Springer-Verlag, New York, 1974.

W14 WONHAM, W.M., "Linear Multivariable Control—A Geometric Approach". Springer Lecture Notes in Economics and Mathematical Systems, Vol. 101. Springer-Verlag, Berlin, 1974.

Y1   YOSIDA, K., "Functional Analysis". Springer-Verlag, Berlin, 1965.

Z1   ZADEH, L.A. and DESOER, C.A., "Linear Systems Theory". McGraw-Hill, New York, 1963.

Further papers

The papers listed below, while not supporting the text directly, will be found useful in developing further the concepts of the book.

CHEN, C.F. and YATES, R.E., Approximating systems with infinite-dimensional state-space. *Internat. J. Systems Sci.* **8**, No. 11, 1299–1311 (1977).

DAVIES, T.V., A review of distributed parameter systems theory. *Bull. Inst. Math. Appl.* **12**, No. 5, 132–138 (1976).

GIBSON, J.S., The Riccati integral equations for optimal control problems on Hilbert spaces. *SIAM J. Control* **17**, No. 4, 537–565 (1979).

KLUVÁNEK, I. AND KNOWLES, Z., "Vector measures and control systems". North Holland Mathematics Studies, Vol. 20. North Holland, Amsterdam; American Elsevier, New York, 1975.

KLUVÁNEK, I. and KNOWLES, G., The bang-bang principle. *In* "Mathematical Control Theory" (Ed. A. Dold and B. Eckmann). Lecture Notes in Mathematics, Vol. 680. Springer-Varlag, Berlin, 1978.

KNOWLES, G., Some problems in the control of distributed systems and their numerical solution. *SIAM J. Control* **17**, No. 1, 5–22 (1979).

KOBAYASHI, T., Some remarks on controllability for distributed parameter systems. *SIAM J. Control* **16**, No. 5, 733–742 (1978).

LEIGH, J.R. and MUVUTI, S.A.P., A practical control algorithm derived through functional analysis. Preprints Internat. Conf. Control and Its Applications, University of Warwick. Publ. IEE (UK). (Scheduled) 1981.

ROLEWICZ, S., "On the control of systems with distributed parameters". (Proc. Internat. Symp., Darmstadt, 1976). Lecture Notes in Math, Vol. 561, pp. 412–420. Springer, Berlin, 1976.

SCHMIDT, E.J.P.G., The "Bang-bang" principle for the time-optimal control problem in boundary control of the heat equation. *SIAM J. Control* **18**, No. 2, 101–120 (1980).

WILSON, D.A. and RUBIO, J.E., Existence of optimal controls for the diffusion equation, *J. Optimization Theory Appl.* **22**, No. 1, 91–101 (1977).

Subject Index

A

Adjoint, 61
Adjoint variable, 113
Attainable set, 104–105

B

Banach–Alaoglu theorem, 47
Banach inverse theorem, 36
Banach space, 19
Banach–Steinhaus theorem, 37
Bang-bang control, 109, 135
Boundary, 5
Boundary point, 5
Bounded inverse theorem, 36
Boundedness, 5

C

Cardinality, 2
Cauchy sequence, 5
Causality, 51, 55
Closed ball, 12
Closed graph theorem, 36
Closed mapping, 35
Closed set, 6, 11
Closure, 6
Compactness, 12–13
   failure of, in infinite dimensional
     spaces, 46
   preservation under bounded linear
     mappings, 13
   weak, 45
   weak*, 45
Complement, 2

Complete space, 5
Continuity,
   strong, 48
   uniform, 48
   weak, 48
Continuous time system, 52
Controllability, 71–76
   distributed systems, 130–132
   time invariant systems, 74–75
   time varying systems, 73–74
   weak, 131–132
Convergence, 5
   in norm, 44
   of operators, 47
   pointwise, 13
   strong, 44, 47
   uniform, 14, 47
   weak, 45, 47
   weak*, 45
Convex set, 6
Cost indices, 79

D

Dense set, 6
Dimension, 3
Direct sum, 7
Discrete time system, 52
Domain, 7
Dual space, 21
   normed, 22
   second normed, 22
Dynamic system, *see* Linear dynamic
   system

157

E

Essential supremum, 19
Euclidean space, 16
Exposed point, 35
Extrema, 86–88
Extreme point, 35

F

Finite dimensional system, 52
Fourier series, 102
Function, 7
  bisective, 7
  continuous, 8
  insective, 7
  inverse, 8
  linear, 8
  locally linear, 8
  operator valued, 48
  surjective, 7
  uniformly continuous, 8
Functional, *see* Linear functional

G

Graph of a mapping, 36
Greatest lower bound, 4
Groupoid, 121

H

Hahn–Banach theorem, 33–35
Hamiltonian, 113
Hilbert space, 27
Hille–Yosida theorem, 124
Hölder's inequality, 37
Hyperplane, 34
  support, 34

I

Image, 8
Infimum, 4
Infinitesimal generator, 122
Infinitesimal operator, 122
Inner product, 26

Inner product space, 26
Input space, 50
Interior point, 5
Interior of a set, 5
Intersection, 2
Intervals, 6
  closed, 6
  open, 6

K

k cube, 7

L

L^p spaces, 18
Least upper bound, 4
Lebesgue integral, 16
Lebesgue measure, 15
Limit,
  on the left, 121
  on the right, 121
  of a sequence, 5
Limit point, 6
Linear dynamic system, 52
Linear functional, 21
  bounded, 21
Linear independence, 3
Linear manifold, 34
Linear space, 3
Linear subspace, 3
Linear variety, 34
Lyapunov's second method, 65–67

M

Manifold, *see* Linear manifold
Mapping, linear, 8
  continuous, 8
  bounded, 8
  of finite rank, 41–42
Maximum, 3
Mazur's theorem, 35
Measure, 15
  inner, 15
  Lebesgue, 15
  outer, 15

Metric, 4
Metric space, 4
Milman and Pettis theorem, 44
Minimum norm control,
    in Banach space, 86–93
    distributed systems, 132–133
    in Hilbert space, 81–86
Minimum time control,
    characterization, 107–112
    distributed systems, 132–135
    existence, 105–106, 134
    problem description, 103, 133
    uniqueness, 106–107, 135
Modal matrix, 60

N

Neighbourhood, 5
Nilpotent matrix, 62
Norm, 4–5
    examples of, on G_T, 94–95
    operator, 9
    product space, 39
Normed space, 5
Null set, 15
Null space, 9

O

Observability, 75–76
One to one correspondence, 8
Open mapping theorem, 36
Open set, 5, 11
Operator, linear, 8
Orthogonality, 26
Orthonormal basis, 27
Orthonormal set, 26
Output mapping, 50
Output space, 50

P

Parallelogram law, 28
Periodic system, 56
Pontryagin maximum principle, 112–116
Product metric space, 7
Product set, 7
Projection theorem, 29

R

Range, 7
Rank of a mapping, 41
Reachable set, 119
Realization, 57–58
    minimal, 58
Reflexive space, 42
Regulated function, 121
Resolvent, 124
Resolvent set, 124
Riemann integral, 17
Riesz representation theorem, 30
Rotund space, 43

S

Schauder basis, 27
Schwarz inequality, 31
Semi-group, 121–124
    transformation, 122
Semi-group property, 51, 54
Sequence, 3
Sequence spaces, 16
Set, 2
    denumerable, 2
    nondenumerable, 2
Smooth space, 44
Space of all bounded linear mappings, 22
Spectrum, 124
Stability, 63–70
    definitions, 64
    for distributed systems, 129–130
    for time-invariant systems, 64
    for time varying systems, 67–70
State space, 50
State transition mapping, 50
Subsequence, 13
Subset, 2
Subspace, see Linear subspace
Support functional, 35
Supremum, 3

T

Time invariance, 51, 56
Time-optimal control, see Minimum
    time control

Topological space, 11
Topological vector space, 11
Topology, 11
  natural, 21
  stronger, 12
  weak, 46
  weaker, 12
Transformation, 122
Transition matrix, 58–61

Uniform convexity, 44
Union, 2
Unique decomposition property, 29
Upper bound, 4

V

Vector space, 3
Volterra operator, 61

U

Uniform boundedness theorem, 37

W

Weierstrass theorem, 46